SMALL STATES IN THE GLOBAL SYSTEM

Small States in the Global System

Analysis and Illustrations from the Case of Iceland

BJÖRN G. ÓLAFSSON

Ashgate

Aldershot • Brookfield USA • Singapore • Sydney

Published by
Ashgate Publishing Limited
Gower House
Croft Road
Aldershot
Hants GU11 3HR
England

Ashgate Publishing Company
Old Post Road
Brookfield
Vermont 05036
USA

British Library Cataloguing in Publication Data
Ólafsson, Björn G.
  Small states in the global system : analysis and
  illustrations from the case of Iceland
  1.States, Small 2.Iceland - Economic conditions - 1945-
  3.Iceland - Politics and government - 20th century
  I.Title
  321'.06'094912

Library of Congress Catalog Card Number: 97-076951

ISBN 1 84014 129 8

Printed in Great Britain by
Antony Rowe Ltd, Chippenham, Wiltshire

Contents

List of tables and figures

Foreword

A few years ago, I complained in print (Karlsson, 1992) that Icelandic social scientists (I was thinking particularly of political scientists and economists) had, on the whole, failed to take seriously enough the special character of very small states and had therefore quite possibly failed, in significant ways, to shed light on the situation of Iceland. As may be seen from the bibliography of the present work, there existed, to be sure, a "small states" literature to which Icelandic social scientists had contributed; and this has thankfully increased. But that literature was, and still is, largely based upon theories, methodologies and mental sets developed for the purpose of studying modern nation-states which are typically much, much larger than the Republic of Iceland. So even this "small states" literature is in some danger of remaining "blind" to certain of the special—and perhaps highly important—ramifications of small size.

I referred, in my (very) little piece, to Galileo Galilei and to the biologist, J. B. S. Haldane, both of whom had pointed forcefully to the fact that things of very different size require very different forms or structures in order to work effectively, to survive, or even to exist in the first place. At least, they argued, this applies to mechanical devices and living organisms—a mouse blown up to the size of an elephant would simply collapse, for example— and so why not to social and political institutions? From this perspective, viewing a small state such as Iceland with the preconceived expectation that it will be, or should be, simply a miniature version of a state like France or Germany (or even of the United States, from which country much of contemporary economic theory derives) could lead one into making fundamental mistakes of both theory and practice.

Will a parliamentary structure adopted from Denmark well suit a country the size of Iceland? Will a central bank on the European model be effective in the regulation of the Icelandic national economy? These were questions that were never seriously raised in Iceland in the course of political debate, or even at the theoretical level by social scientists. The structures in question were simply taken up in the belief that they would function in Iceland in the same way that they were known (or thought) to function in the larger European states. But do they so function? And if not, why not? And what alternative structures might better fit a small—and semi-isolated—island state? Certain small states, such as Switzerland and Jersey, have developed structural institutions which seem better suited to their size than the "borrowed" institutions which we find in Iceland. And I wanted Icelandic economists and political scientists to forget their European and American prejudices long enough to investigate these questions with the possibility in mind (among others) that, as Galileo and Haldane emphasised, structure and size are intimately related and need to go hand in hand.

I do not know how much of the inspiration for the present work came from my little article. Surely not much (although the article is quoted briefly, and to good effect, in chapter 2), for Björn G. Ólafsson's *Small States in the Global System* was already well under way when my article appeared. But the book certainly represents a move in the direction I desired. I do not pretend to be well enough versed in economics to pass a professional judgement on the merits of Björn's work; its level of sophistication is far beyond me. But the book is, to its credit, straightforwardly written and accessible to the layman. And one can see clearly that an attempt has been made throughout this work not to assume that small states are merely smaller versions of larger states, but to look at a great many fundamental economic questions with unblinkered eyes. Thus, the book commences with a discussion of the concept of the "small state" and of the theory of optimal size. A chapter is devoted to the special economic characteristics of small states, and studies which seemed to show that small-state economies did not exhibit any special characteristics are effectively criticised. Monetary and trade questions, as they apply to small states, are also examined in a refreshingly unprejudiced way.

Björn's main conclusion, which is that small states are not suboptimal units in the global system, will surprise many and will undoubtedly provoke controversy among economists. But in my judgement the interest of this work lies as much in its fresh and open approach as in its conclusions, important as those conclusions may be. It is indeed a ground-breaking effort in more than one respect.

Mikael M. Karlsson
University of Iceland

Preface

This study contains my doctoral thesis submitted to the University of Exeter in 1995. The original text of the thesis is mostly unchanged but the appendix, which contained statistical background material, is excluded. This should not be a problem because most of the information from the appendix is summarised in the tables and graphs in the main text with references to the original sources. Furthermore up to date statistical material on small states is now published on the internet or otherwise easily obtainable in printed or electronic form from a variety of sources.

I wish here to repeat my thanks to those who read early drafts of the thesis or supported the study in other ways: Jónas Ólafsson, Vigdís Wangchao Bóasson, Emil Bóasson, Sigurður Guðmundsson, Guðmundur Malmquist, Helga Finnsdóttir and Ólafur Björnsson. Lilja Karlsdóttir assisted in the drawings of diagrams and the layout of the final print. My parents Ólafur Björnsson and Guðrún Aradóttir, my wife Helga Finnsdóttir and my children, Ólafur Darri and Guðrún Ása, have given support and encouragement all the time. My supervisor at the University of Exeter, Jeffrey Stanyer, provided invaluable advice and comments on the thesis. Finally I thank Mikael M. Karlsson for the foreword and the editorial staff at **Ashgate,** in particular Sonia Hubbard, Amanda Richardson, Kate Hargreave and Pauline Beavers. All errors and omissions are the sole responsibility of the author.

The international system is constantly changing. Small states have the potentiality to be flexible and adjust rapidly to new economic opportunities and political challenges. Advances in communications such as the internet have effectively reduced distances and enabled small island states to integrate more easily into the world community. The software industry offers many new employment opportunities even in remote places. Other technical breakthroughs such as microscopic nanotechnics are likely to effectively increase the resource base of small states. New opportunities also exist in the tourist industry such as whale watching and sea-angling. On the international arena small states must actively participate in the solution of global environmental problems as well as problems related to the management of fishing stocks and other natural resources. For the small state, the key to success in the global system is education and investment in human capital. I hope that this study will stimulate further research into the problems and potentialities of small states.

1 Introduction and definition of a small state

This study examines the economic and political status of small states with a special reference to Iceland. The questions raised are related to the economic viability of small states and their political strength in a changing world. No attempt will be made to find answers that apply unconditionally to all situations or to all small states. General theories are used to solve the problems that arise but the focus will be on Iceland as a special case. The study is by nature interdisciplinary and combines ideas and material from political, geographical, legal and economic sources. Elements of a theory of *the optimum size of states* are developed as a framework for analysis of small states.

The nature of the problem

Pessimism was expressed about the abilities of small states to function independently in the world community when the number of small states began to increase in the nineteen sixties and nineteen seventies as a result of the application of the principle of self-determination to former colonies by the United Nations.[1] Small states were considered suboptimal and therefore vulnerable to political and economic pressures of all kinds that arise in international relations and international trade.

In the first instance the problem was and is seen as an economic problem. The question of the economic disadvantages arising from smallness was an important topic at the pioneering Lisbon conference on small states in 1957 but at the time only a handful of independent states with populations of less than one million existed.[2] Since then several studies have analysed economic problems of small states. A general result of the analysis is that economic vulnerability is a consequence of undiversified production and trade. The small state obtains the benefits of international trade by specialising in the production of few commodities in which it has comparative advantage. Increased specialisation makes the small state economy more vulnerable to fluctuations in export prices and more dependent on a few market outlets. Many large states have a similar problem, especially developing states, but the small state faces a dilemma between the need to diversify and the need to specialise. Diversification is needed for the economy to make it less dependent on a few overseas markets and a few export products. Specialisation is needed to obtain the benefits of trade. This study shows that despite the fears expressed at the Lisbon conference and subsequently in other studies, there are no insuperable economic disadvantages resulting from smallness that prevent small states from obtaining high standards of living. Fredrich E. Schumacher (1989:76) is probably correct when he writes: 'How can one talk about the economics of small independent countries? How can one discuss a problem that is a non-problem? There is no such thing as the viability of states or of nations, there is only a problem of viability of people [...].'

The second aspect of vulnerability of small states was and is political. Questions of diplomatic power and military strength are an important part of classical small state studies. Several works on small states, both old and recent, have treated this problem. Small states in such studies are mostly defined to include states that have a population of several millions. For instance David Vital (1967) uses an upper limit of 10–15 million and 20–30 million for developed and developing countries respectively. Michael Handel (1990) in his theory of *weak states* includes in the class of weak states countries with populations up to 40 million. Questions of political power on a global scale are irrelevant for small island states with populations below one million in most cases. The existence of such small states therefore depends on an international system which is based on law and order as well as international cooperation. Participation in international institutions and diplomatic activity is costly and small states must therefore limit their diplomatic activities to a few key areas. The small state needs a stable international system but has limited means to influence developments in that system. This study shows that small independent states especially islands with a large *Exclusive Economic Zone* (EEZ) have, nevertheless, a relatively strong political position in the global system.

If pessimism as to the viability of small states is based on matters of culture, education and administrative capability then it is unfounded. Iceland has clearly demonstrated that the limit of smallness in these matters is well below 250,000 inhabitants. Iceland has several institutions providing higher education. The largest one, University of Iceland, has around 5,500

2

students (1995). It also has a symphony orchestra, several theatres and an opera. Although postgraduate courses are only offered in a limited number of fields, Icelandic students have been able to obtain specialised education in various institutions around the world providing the society with a full range of specialist services and research. Lack of skilled personnel or the ability to offer rich cultural life is therefore not a problem of small societies except perhaps as a consequence of economic underdevelopment and a lack of investment in human capital.

Definitional problems

A commonly accepted definition of a small state is not available. The definition of a small state is usually the first topic discussed by authors who write about small states. No lengthy methodological discussion of this problem is necessary here. This study does not deal with the problem of finding a fundamental characteristic that can be used to explain other characteristics of all small states. Population is used as a measure of size and the sample of small states for statistical purposes is limited almost entirely to countries with a population below one million. This is a matter of convenience as a limit of 1.2 or 1.5 million would probably not make much difference. Lack of a commonly accepted definition of a small state is not a hindrance for research on small states.

The definition of *real independence* and the related concept of an optimum size is an important problem. A formal recognition of sovereignty by the international community is not equivalent to real independence. An independent small state may be dependent on another state or states, often the former colonial power, in matters of foreign policy, administration and economic aid. An exact definition of independence is not provided but an *independence/dependence continuum* is used to clarify the status of small states in relation to other states. A definition of the optimum size of states is not given beforehand but the concept will be clarified in this study. In a state of optimum size the preferences of the citizens should influence as much as possible the policy of the government (*citizen effectiveness*) and the state as an organisation should have as much capacity as possible to fulfil the wants of the citizens (*system capacity*).[3] A complication arises because the state has two faces, internal and external. The state can be regarded as an actor in international relations where the state reflects the collective interests of its citizens. Optimum size in this context relates to the capacity or power of the state. The power of a state is created by many socio-economic-geographical factors such as the size of the population, the size of the economy and the area. On the other hand the state manages several functions for the citizens. Optimality is then related to the effectiveness of the state as an (democratic) organisation. This study is mainly concerned with the relation between capacity and the size of the state. It is simply assumed at the beginning that citizens in small states have more influence on the government than citizens in larger states. Lack of system capacity (if any) is regarded as the price to pay for more democracy or more direct influence on the government.

Theoretical bases

Is there an optimal size of states? Is the small state suboptimal as a political and economic unit in the global system? Several disciplines contribute to the answer to these questions. Most studies of small states have either discussed the political aspect or the economic aspect of the size of states but not both at the same level of analysis. The problem is that it is often difficult to draw the dividing line between economic causes and political causes in international relations.[4] By answering the question of optimal size of states an integrated view of the problems and prospects of small states can possibly be constructed.

The contribution of political science

An analysis of the optimum size of all governmental units involves federalism, local government, secession movements and special status regions.[5] Although the idea of optimality has a certain 'utopian' flavour, it helps to clarify the problems that arise as the size of the state increases. In large states the lines of communication from the central government to the individual citizen are very long. Large states may be slower to respond to changes and opportunities in the political and economical environment than small states. But small states are suboptimal as an international unit in many ways and may lack political power and economic capacity. To discuss the idea of an optimal size of states, concepts from systems theory are used in conjunction with ideas from information and democratic theory.

The discussion of independence of small states and related topics refers to the theory of weak states, which is discussed briefly and applied to the problem of real independence. An additional input is *the Olsonian theory* of groups and organisations.

International law appears as an element of political science. Discussion on the implication and importance of the Exclusive Economic Zone has so far mostly belonged to the field of international law. The political implications of the EEZ have been largely neglected in works on small states although the importance of fishing and fish-processing industries has been discussed.[6] It is clear that the administration and intelligent utilisation of

4

the EEZ will be one of the most important subjects of small state politics and economics in the future.

The contribution of economics

The theory of the *second best optimum* is used to investigate how optimum combinations of citizen effectiveness and systems capacity are affected by the size of the population or other constraints that are significant in the analysis of the optimum size of states.

International trade and trade characteristics of small states are examined in the light of *the Heckscher-Ohlin theory* of the basis of trade. A more general approach, which puts the development potential of small states into international perspective, is to analyse small states within the context of the theory of *intra-industry trade*. One of the main characteristics of small states is that their foreign trade is of inter-industry type and that the conditions for intra-industry trade are mostly lacking. The index of intra-industry trade of small states is calculated although such measurement involves several unresolved methodological problems. This analysis is combined with the theory of *customs' unions* to estimate possible economic benefits of regional integration.

International trade and economic development come together when problems of monetary policy are considered. Is a small state such as Iceland *an optimum currency area*? The question of fixed versus flexible exchange rates and the effects of inflation leads to consideration of the role of money as a firm measure of values and an instrument of exchange.[7]

The contribution of regional science

Regional science has much to offer in the study of small states which have many characteristics of peripheral regions in larger states. Economic growth is unequally distributed in geographical space and small states are affected by 'backwash' and 'spread' effects of fast growing regions. *Locational* theory which includes the theory of *growth poles* or *growth centres* helps to clarify the relation between growth in economic space and growth in geographical space. It is also a necessary complement to the theory of international trade. Inside small island states the regional problem is important, especially from the point of view of keeping a balance in the settlement structure without hindering urban development which is necessary for the growth of service industries. Seen as a problem of distribution the regional problem easily gives rise to political conflicts. Discussion of the *area factor* includes regional problems and the effects of an Exclusive Economic Zone on the size of small island states.

Areas that are small states face the choice between independence and various degrees or forms of integration. This situation leads to the question whether there are any disadvantages associated with independence that make some form of (regional) integration a better option for small states. Various types of regional associations are developing around the world. The practical question for small states is what kind of (regional) integration leads to optimum economic and political capacity. For Iceland the question is what economical and political benefits will be obtained by entering the European Union (EU).[8] *A priori* small states are generally more likely to gain from a general liberalisation of trade through the General Agreement on Tariffs and Trade (GATT) regime than by entering a customs union.[9] This applies especially to small states in remote locations where regional integration is not practical. Small states are, however, inevitably spectators of the GATT negotiations, and may end either as free riders or as victims.

In brief, analysis of the concept of real independence and the optimum size of states indicates that the independent state has certain advantages over semi-independent or autonomous regions within larger states. Regional integration may therefore not be the solution of the problem of suboptimality.

The unequal world of small states

States and territories of the world differ in terms of population and geographical size. On one hand there are the large states of China and India with over one third of the world's total population or 1,156 million and 850 million inhabitants respectively, and on the other hand there are some independent states with populations under ten thousand.

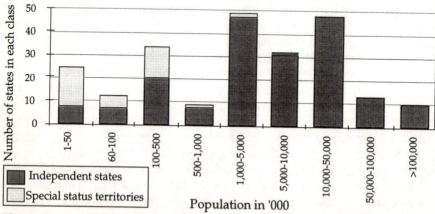

Figure 1.1   Distribution of countries by population size 1991
Source: United Nations (1993).

The distribution of states by population is shown in figure 1.1. The states have been grouped into nine classes.[10]

The meaning of the term independence will be discussed more fully later on but the dividing line is unclear between the group of independent states and other special status territories which have not been formally declared fully independent. Countries such as the Cook Islands are self-governing but in free association with another country, in this case the United States. Some states have 'home-rule' in the manner of Greenland and the Faeroe Islands but many former colonies of France have obtained the status of overseas departments or territories and are at the same time considered an integral part of France. The number of independent small states with populations under one million could therefore be increased if all the special status territories obtained full independence.

Data has been collected for 80 small countries and territories with populations less than one million.[11] Of these 47 have an area less than 1,000 square km. Except for Greenland, Iceland, French Guiana, Suriname and Western Sahara most of the small states possess a limited land area which often means that they have a limited range of both natural resources and arable land. Six of the small states are landlocked, namely Andorra, Liechtenstein, Luxembourg, San Marino, Swaziland and The Vatican City. Most of the others are islands or archipelagos. Geographically it means that many of them are remote and isolated from large markets but it also means that they can establish a 200 mile Exclusive Economic Zone which effectively enlarges the area over which they have sovereignty.

Table 1.1

Area and population of countries

Population in millions	Number of countries	Countries with EEZ	Total population million	Share of population	Total area '000 sq. km	Share of area	Area + EEZ '000 sq. km	Share of area + EEZ
<1.0	80	30	17	0.3%	3,324	2.5%	15,181	6.0%
1.0–4.9	49	27	147	2.7%	11,240	8.3%	17,768	7.0%
5.0–9.9	32	15	229	4.3%	9,339	6.9%	10,660	4.2%
10.0–49.9	48	37	974	18.1%	49,976	36.8%	94,781	37.7%
50.0–99.9	13	12	803	14.9%	9,811	7.2%	11,372	4.5%
>100.0	10	10	3,218	59.7%	52,006	38.3%	101,914	40.5%
Total	232	131	5,389	100%	135,697	100%	251,676	100%

Sources: United Nations (1993), United Nations (1992), National Geographic (1993), Hoffman (1992), Attard (1987), Seðlabanki Íslands [The Central Bank of Iceland] (1993).

The class of states with populations less than one million has only a 0.3% share of total population but controls 6.0% of total area if the EEZ is included as can be seen in table 1.1. The potential share of small states would increase if the EEZ of all dependencies were shown separately but often the EEZ of overseas territories is included with the ruling state. An EEZ has also not been established for all countries. It is clear from this table that the small states and small territories control a significant part of total land and sea area although they have a negligible share of the world's population.

Definition of a small state and the measure of size

Population is used here as a measure of size but other variables such as area and Gross National Product (GNP) have been used in combination or put alongside population as independent variables.[12] The definition of a small state is derived from a survey of their overall statistical distribution. Small states are presumed to be at the lower end of each scale and the choice of a cutting point is arbitrary (to some extent). Climate, culture and history are also important explanatory variables in the comparative study of small states but such factors are difficult or impossible to quantify and require a different methodological approach from that used in this study.[13]

It is probably not possible to find a definition of size that is fundamental in explaining all characteristics of small states in general. This situation might lead one to doubt that the small state can become a useful unit of analysis in international relations. An exact definition is, however, not necessary to study various aspects of states and to compare them in terms of size. The meaning of smallness as a theoretical term will be clarified through the study of various characteristics and problems of small states.[14]

Population as a measure of size

The most important measure of size is population. The size of the population has many obvious consequences for social structure and processes. Firstly it determines to a considerable extent the size of the internal market before the foreign trade factor comes into operation. Secondly it determines the possible internal division of labour and the degree of internal specialisation within a country whether the activities are of an economic or social or political nature. Thirdly it is the human resources or human capital of the country.[15]

The size of the population is not necessarily stable but large changes in a short period of time are unlikely. Only severe political unrest leading to the breakdown of states or large scale natural catastrophes can alter the relative positions of states in terms of populations. In other words states tend to remain at the same relative position on the population scale over a period of time.

Area as a measure of size

The second measure of size is the geographical scale or area. Just as the population is 'human resources' so do territories have varying natural resources which are not necessarily related to size as such. Small population does not always imply a small area (otherwise it would not be necessary to consider area as a size variable). Many countries are large in geographical area with a small population, for instance, Greenland. In such cases a great deal of the country is often uninhabitable at the present level of technology. Lloyd and Sundrum (1982:20–21) suggest that habitable or cultivated land would be a more pertinent index of geographical size than total area but this does not solve all the problems.

Growth of the population was mostly dependent on the availability of arable land when land-based activities were dominant in supplying the means of existence. Countries that have a reasonably long history of settlement have probably utilised arable land in similar proportions at least until industrialisation started. The term arable land is, however, not clearly defined and its size is not constant. New resources can be found or technical progress might make land, which is unusable now, fertile. Global climatic changes may also affect the fertility and size of arable land. Although it is often not clear to what extent the sea can be included as arable land, rich fishing grounds can be just as valuable as fertile soil. Many small states become much larger in terms of geographical size if their Exclusive Economic Zone is included as arable land.

Geographical size has important consequences for the economic and political status of states although there is no direct relationship between geographical size and the size of the population. In the case of a country with a small population dispersed over a large area as is the case with Iceland and many archipelagos, certain regional problems arise which are directly related to this combination of geographical factors and a small population. For this reason it could be useful for analytical purposes to define a small state in terms of population size and inverse geographical size, that is, the larger the area+EEZ and the smaller the population the smaller or weaker the state. Landlocked states of a limited geograpical size such as Luxembourg and Andorra can be regarded as small regions rather than small states for analytical purposes. The area factor is discussed in greater detail later but geographical size is clearly an important measure of size although it is not useful as an exclusive or sole measure.

The third candidate as a supplementary size variable is Gross National Product. GNP is an indicator of the size of the internal market. As a measure of total available resources, GNP can serve for example as an indicator of military potential that is important in the study of political power. If there are large fixed costs of being a state then the question of how much of GNP is needed for these becomes important.

One problem with GNP as a measure of size is how changeable it is. Rapid economic growth, based on a discovery of oil deposits for example, can change the relative position of a state within a short period of time. Great changes in the relative position of states, especially at the lower end of the size scale, might occur if size is measured in variables that are dependent upon policy, subjective factors or economic factors. Classification of states in terms of such variables will therefore be unstable and no balance between relative and absolute positions. Another problem with GNP as an indicator of size is that the degree of economic development is a major determinant of its size. Therefore it is just as important as a factor to be studied as a consequence of smallness rather than a simple measure of smallness. It therefore seems best to regard Gross National Product as a dependent variable rather than an independent measure of size.

Composite measures of size

After discussing the various measures of size, Lloyd and Sundrum (1982:20) find that there is no systematic relationship between population size and other measures of size. 'The ranking of countries and the partitioning of the set of all countries into the two groups of "small" and "large" will differ according to the criterion adopted.' They then discuss the possibility of constructing a composite index of size out of statistical methods. But the use of 'principal component analysis' and 'discriminant function analysis' for this purpose did not produce useful results.

Bimal Jalan (1982:44) constructed a size index out of a sample of 111 countries, where equal weights are given to Gross National Product, area and population as independent measures of size. Using a median value of this index as a cut off point for small states he suggested that populations below 5 million (in 1977), an area of 25,000 sq. km and GNP of US$3 billion should be the boundary for the classification of small states. In constructing his index of smallness Jalan (1982:42) is actually using population as a proxy for labour force, total national income as a proxy for capital stock and area as a proxy for arable land. His underlying hypothesis is that differences in economic structures and economic performance due to the size factor are likely to be due to differences in the resource base of countries as represented by the size of their capital, human and natural resources.

Many problems arise when the three very different variables GNP, population and size are added together to create an index of size. Jalan's use of arable area in constructing his index is subject to the above mentioned problems of measuring arable area and he did not take the existence of EEZ into account. GNP is also an unstable variable because it is the result of economic development. There is no reason to believe that adding these variables will reduce the problems connected with the use of each of them as a sole measure of size.

Differentiating small states

The focus of this study is the case of Iceland; its population size of around 265,000 (1993) will be used as a reference point. The geographical situation of many small states that are islands is close to Iceland and creates similar regional problems. Other factors such as climate and history are very different from that of Iceland. Despite such differences it seems that the case of Iceland can safely be generalised to island states with a population at least four times the size of the Icelandic one. If the geographical situation is very dissimilar the area factor might affect the conclusions in some cases.

How far down the scale of size is it necessary to go? Do states of 10,000 inhabitants warrant special treatment in this study? This class of states is probably characterised by the absence of governmental apparatus sufficient for participation in international organisations or other activities related to foreign affairs. The lower end of the scale will therefore not be of interest at this stage of the analysis. The upper limit is what matters to delimit the class of small states from large states. In existing studies the upper limit of the class of small states has ranged from 300,000 or less to 15 million or more.[16] In this study the class of small states in statistical analysis is limited to those of population around one million. This class of states will be used for international comparison in statistical analysis. More substantial aspects of the concept of smallness will, hopefully, emerge as this study progresses.

Notes

1   A study by the United Nations Institute for Training and Research (Jacques Rapaport et al., 1971) is specially concerned with the role and participation of very small states in international affairs, see also Burton Benedict ed. (1967) and Patricia W. Blair (1967).

2   For papers and proceedings of this conference see E. A. G. Robinson (1960).

3   Problems related to social choice and majority voting are ignored here.

4   Handel (1990:217) says that most studies of weak states in international relations are strictly political and military and the economic studies that exists rarely include political-military analysis. He finds this separation between the economical and the political and military spheres to be unnatural as they are closely related in reality.

5   The question of the optimum size of governmental units is practical in the case of local government. In Iceland, for example, a general referendum in 1993 on new proposals for reducing the number of districts from about two hundred to between forty and fifty was not successful. Only a minority of rural districts agreed to merge with larger ones. Enlargement of administrative units was to be accompanied by the transfer of several functions from the central government to the new local governments. The reasons for failure were many but a free riding problem was doubtless one of them.

6   Antony J. Dolman (1985) discusses some implications of the EEZ for small states.

7   The measure of money is a problem because in the end the amount of money measures the level of creditworthiness or credibility in the society.

8   Iceland participated, with other EFTA members, in the creation of the European Economic Area (EEA). Effective from January 1994.

9   The World Trade Organisation (WTO) replaced GATT in 1995.

10   This diagram shows the distribution of independent states and the distribution of independent states plus territories that are not fully independent or of a colonial status.

11   Collecting basic data on states such as population, area and trade was, surprisingly, not a simple job. There are inconsistencies in the data from different sources and the nature of the discrepancies is difficult to discover. Statistics from the official publications of the United Nations are used throughout this study (where possible). Data on the area of the Exclusive Economic Zone is probably subject to large margins of errors. The limit of the EEZ is in some cases still a matter of dispute.

12   It makes no difference in this context whether one uses Gross Domestic Product (GDP) or Gross National Income (GNI) instead of Gross National Product.

13   Instead of using population, GDP or area to classify states Handel (1990) proposes the construction of two ideal types of states 'weak states' and 'strong states'. His ideal weak state has many characteristics of small states when size is measured with those variables.

14   Terms that require for a specification of their meaning the context of

the whole set of sentences in which they appear have 'systemic meaning'. See Abraham Kaplan (1984:63–65).

15   'The size of population is very important as a measure of consumers and a measure of labour force. Therefore whatever index of size we adopt, it will have to give considerable weight to population size. The question is whether it needs to be supplemented by any other measure of size to identify the group of countries whose development prospects are seriously affected one way or the other by their smallness' (Peter J. Lloyd and R. M. Sundrum, 1982:18).

16   See Philippe Hein (1985, appendix A) for an overview of different views on smallness.

2   The optimum size of states

Discussion of the advantages and disadvantages resulting from a small size of states in the global system suggests a theory of an optimum size of states. In a state of optimum size the preferences of the citizens should influence as much as possible the policy of the government and the state as an organisation should have as much capacity as possible to fulfil the wants of the citizens. In a small state the population is the main constraint on the capacity of the state but other constraints are also operative. A single optimum size of states in terms of population that satisfies all criteria and constraints can probably not be found. To make some progress it is necessary to divide the state into subsystems. The main interest here concerns the political system and the economic system. It turns out that the citizen of a small state has a better possibility to influence decision making than a citizen in a large state. Small states may, however, lack economic and political capacity to fulfil the wants of the citizens and solve the problems that arise in international relations. There is a trade off between system capacity and citizen effectiveness. A new application of the theory of second best shows that although the population constraint prevents an optimum situation departure from an optimum in several subsystems may be the second best option for small states.

The concept of the optimum size of states

The concept of the optimum size of states is beset with difficulties and even if these difficulties could be overcome it is doubtful that an exact optimum size of states in terms of population or some other measure of size exists. Historically the spectrum of the desirable or the optimum size of states in various ideologies has ranged from the small city state or polis of the Greeks to the idea of a world government that was put forward in the first years after World War II as a solution to the armament race and the threat of the atomic bomb.[1]

The small state is not a suboptimal unit in the global system if it can be shown that a small state can respond to the preferences of its citizens and solve problems that arise in a satisfactory manner. Today many states with populations less than 20 million seem to offer their citizens more welfare and fewer social problems than larger states. In this class are the Scandinavian countries. In this class are also states such as Iceland and Luxembourg with populations less than half a million. Sweden for example has a highly developed economy containing a large sector of modern manufacturing industries. In international trade Sweden seems to have few disadvantages. This observation applies also to the other Scandinavian countries and Switzerland. Experience shows that the lower limit of the optimum size of a state is below 20 million and probably down to 200,000.

Optimality can sometimes be seen as a combination of maximum and minimum of some social situations but in other cases increases above some minimum values are irrelevant or of no causal significance. The minimum or 'threshold' value is important in many cases. When important social processes such as public services or public utilities are indivisible up to some level of population size, small states below that threshold are suboptimal. Problems of indivisibilities and unattainable threshold values arise easily where the settlement structure is scattered as in peripheral regions in large states and in small states containing archipelagos. The reason is that the basic rights of the individual are indivisible and independent of whether he or she belongs to a small or a large state. Every individual of a large or small state must also have access to basic education, food, shelter and health services. The problem of small states is traditionally thought to be their inability to reach a minimum level of political power and economic growth to secure the basic rights and needs of the citizens. In these areas it is obviously not possible for small states to reach maximum capacity. The question is whether minimum levels of welfare in small states require unrealistic or unattainable levels of measures such as national income per head. This is not the case as the example of Iceland shows.

States and subsystems

A theory of the optimum size of states requires a simplified conception of the state. For the discussion of optimality it is convenient to regard the state as a system composed of various subsystems. Important subsystems are the economic system, the political system and the cultural system. In an ideal nation-state the boundaries of the subsystems coincide with the geographical boundaries of the state. In practice it is difficult to separate the various subsystems such as the economic system from the political system or the political system from the cultural system. The political process, especially, is closely connected to economic issues and the political and economic systems overlap to a large extent. Many states, especially small states, do not compete in global power politics. For those states international relations are mostly concerned with economic matters although strategic and cultural issues are in no way unimportant. David Easton (1965:57) identifies the political system '[...] as a set of interactions, abstracted from the totality of social behaviour, through which values are authoritatively allocated for a society'. In the case of Easton's theory the political system as such has no size dimension and no unit of analysis comparable to money in economics or bits of information in information theory. Yet his definition is useful to distinguish political functions or actions of the state from functions related to other processes such as market activities.

The political system is different from other subsystems as its role, *inter alia*, is to decide whether it is possible or beneficial to enlarge or diminish the size of various subsystems including the size of itself. Such decisions are either permanent or reversible. Participation in regional integration can be practically irreversible, although it can formally be reversible. Irreversible decisions that aim to enlarge the boundaries of some subsystems may therefore lead to a loss of real independence.

Further complications arise for the following reasons: First, no optimum seems to exist for some subsystems. This happens when the economic system of a state is totally integrated in a larger economic union with no natural barriers. Luxembourg is an example of a state which for many purposes can be regarded as a region in a larger state and therefore its small size becomes irrelevant from an economic point of view. Second, some social processes are unaffected by the size of the state. An example is the output of the law making process such as the rights of the individual that should not be affected by the size of the state. Third, the optimum size of various subsystems may not coincide to produce a grand total optimality of the size of a state. Fourth, there are conflicts between the various optimum criteria as will be shown in the next section.

Despite these difficulties the state will be seen here as composed of subsystems that can be separated for analytical purposes. It is useful to find how each subsystem deviates from the optimum size of such systems in general. The optimum size of a state will then be some (optimum) combination of several subsystems each of an optimum size. The two most important subsystems to be discussed here are the political system and the economical system.

In a state of an optimum size the preferences of the citizens should influence as much as possible the policy of the government or the actions of the state and the state as an organisation should have as much capacity as possible to fulfil the wants of the citizens. Similar, but more detailed criteria of optimality have been advanced by Robert A. Dahl and Edward R. Tufte (1974:20). The ideal (democratic) polity, whether it is located in the city-state or the nation-state satisfies at least two criteria:

1) The criterion of citizen effectiveness (citizens acting responsibly and competently fully control the decisions of the polity).

2) The criterion of system capacity (the polity has the capacity to respond fully to the collective preferences of its citizens).

They then discuss how the requirements for meeting these criteria have evolved in political theory. According to the classical tradition of democratic thought the requirements for meeting the two criteria were interpreted in the following way:

Citizen effectiveness. (1) In order for citizens fully to control the decisions of the polity, they must participate directly in making those decisions. (2) In order to participate directly in making decisions, the number of citizens must be very small.

System capacity. If the polity is to have the capacity to respond fully to its citizens, it must be completely autonomous or sovereign. (Dahl and Tufte, 1974:21)

In the theory of the nation-state the requirements for satisfying the optimum criteria were changed:

System capacity. (1) Only the nation-state has the capacity to respond fully to collective preferences. (2) Therefore the nation-state (but not smaller units) should be completely autonomous.

Citizen effectiveness. (1) Nation-states are too large for citizens to participate directly in all or even most decisions. (2) Therefore, citizens must be able to participate in decisions at least indirectly, typically by electing representatives or delegates to decision-making offices. (Dahl and Tufte, 1974:22)

They mention the third possibility which applies when the nation-state becomes inadequate to solve the problems and is gradually transformed into some sort of local government within a larger international system:

> *System capacity.* (1) Only a complex polity consisting of a number of interrelated political units above and below the level of the nation-state has the capacity to respond adequately to the collective preferences of citizens of nation-states. (2) Therefore units at and below the level of the nation-state should not be wholly autonomous.

> *Citizen effectiveness.* (1) No single method for maximizing citizen effectiveness within each of these units is best. In some, direct participation is best; in some the election of representatives; in others, more indirect methods, including delegation of authority to officials appointed by national governments. (2) Therefore, citizens must be able to participate in decisions at least indirectly, typically by electing representatives or delegates to decision-making offices. (Dahl and Tufte, 1974:25)

Dahl and Tufte (1974:108–109) state that there is a trade off between citizen effectiveness and system capacity as small systems are too costly for many purposes. The criterion of system capacity implies that in the present world there is no single optimum size for democratic polities.

The problems related to the concept of the optimum size of states have not been solved although the above conclusion is accepted. The question remains whether the small state has sufficient capacity to fulfil the wants of the citizens at the same level as larger states can do. This applies in a second best world where all the conditions and criteria for the optimum size of states can never be fulfilled. Trade off between citizen effectiveness and system capacity and the need to adjust the boundaries of the political system to cope with different problems are best treated as constraints that prevent the attainment of the optimum size of states.

The question arises whether the two optimum criteria can also be used for the various subsystems. The criterion of system capacity is clearly a necessary criterion for all subsystems. Any subsystem that cannot respond to the preferences of the citizens is not viable. This criterion is, however, not sufficient. The preferences of citizens are often in conflict and the capacity to respond to the preferences of the citizens is potential until the political process resolves the conflicts that inevitably arise. The criterion of citizen effectiveness is therefore introduced as another necessary condition. These two criteria may not provide both necessary and sufficient conditions for an optimum in all subsystems but they are useful alongside other conditions and constraints that are discussed later.

In the light of the above it emerges that there are conflicts between optimum criteria that make the attainment of the optimum size of states highly unlikely and it may also turn out that a departure from the optimum in many or all subsystems may be the second best solution.

> The general theorem for the second best optimum states that if there is introduced into a general equilibrium system a constraint which prevents the attainment of one of the Paretian conditions, the other Paretian conditions, although still attainable, are, in general, no longer desirable. In other words, given that one of the Paretian optimum conditions cannot be fulfilled, then an optimum situation can be achieved only by departing from all the other conditions. The optimum situation finally attained may be termed a second best optimum because it is achieved subject to a constraint which, by definition, prevents the attainment of an optimum. (R. G. Lipsey and K. Lancaster, 1973:144)

A corollary from this theorem is that there is no *a priori* way to judge between various situations in which some of the optimum conditions are fulfilled while others are not. A situation in which a larger number of optimum conditions are fulfilled is neither necessarily nor likely to be superior to a situation in which fewer optimum conditions are fulfilled. 'The general theorem of the second best states that if one of the optimum conditions cannot be fulfilled a second best optimum situation is achieved only by departing from all other optimum conditions' (Lipsey and Lancaster, 1973:145).

The general theorem of the second best is derived from a mathematical theorem. First it is shown how an optimum can be defined as a problem in maximising a function of a number of variables subject to a constraint. Then if an additional constraint is introduced that prevents one of the proportionality conditions for an optimum, the solution of the new maximising problem leads to new optimum conditions which can all be different from the conditions obtained in the solution without the additional constraint. Therefore if all the conditions required to maximise welfare or reach the Pareto optimum cannot be satisfied, one does not necessarily reach a second best position by trying to satisfy as many of these conditions as possible.

It is not necessary to limit the result to the optimisation (in the Pareto sense) of allocation of economic values only.[2] This theorem applies in all cases where a problem of allocation can be approximated by a function that is to be maximised with given constraints that cannot all be fulfilled.

Therefore the theory of the second best could also be applied to 'authoritative allocation of values' in the political system. However these concepts of optimality apply only to the optimal distribution of values and do not explicitly refer to the optimal size of the economic or political system.

A graphical demonstration of the second best theory when applied to the size of states is given in figure 2.1.

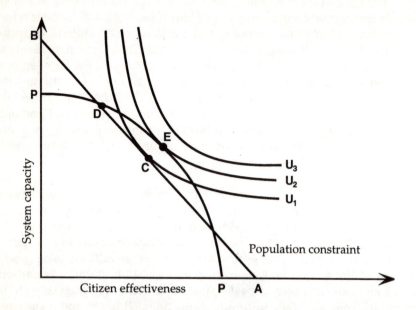

Figure 2.1   **Graphical demonstration of the theory of the second best**

Source: Adopted from Walter Nicholsson (1978:572). The original diagram shows the familiar production possibility or transformation curve for two goods but the variables have been renamed as system capacity and citizen effectiveness.

The shape of the curve PP is determined by the various factors or constraints that affect system capacity and citizen effectiveness. The points E and D on the curve PP represent two possibilities of efficient or optimum combinations of system capacity and citizen effectiveness assuming a trade off between these two. The constraints are for example productive capacity, military strength and information processing efficiency. These constraints will be discussed more fully below. At the end points of PP there is either complete anarchy or complete dictatorship. The U curves represent collective preferences for the kind of state the citizens prefer or social contract curves. The point E represents the optimum size of a state if the size of the

20

population is not relevant for system capacity or citizen effectiveness. The second best point C is preferred to the efficient point D on the optimum curve if a population constraint AB is introduced that prevents the attainment of the optimum.

The second best theorem suggests that if a state is too small to include some social subsystem of an optimal size this may not necessarily make the state unable to obtain satisfactory overall position compared to larger states. Iceland and Luxembourg have a very small home market and are clearly suboptimal in terms of military power. Their citizens enjoy however a higher level of welfare than is the case in most larger states. An optimum size of states that fulfils criteria for optimality in all subsystems is probably impossible. In what follows the conditions for optimality will normally be those of a 'second best' nature but it will also emerge that the population constraint is not a serious constraint on the optimal size of states.

Determinants of citizen effectiveness

Many factors affect citizens efficiency such as participation in voting and sense of effectiveness, communication to leaders and the level of competition in politics. Dahl and Tufte (1974:108) found a trade off between at least two different aspects of citizen effectiveness, cost of dissent and cost of communication and participation. As the number of citizens increases the cost of participation and communication with leaders increases but the cost of dissent declines. Decline in the cost of dissent as the size of the state increases means that there is a greater chance to find an ally in dissent and an opportunity to participate in an organisation with interests and policies that deviate from the views of the majority.

Small states both in terms of population and geographical size were generally thought to be ideal in classical writings before the advent of the large nation-states.[3] An optimum size of a state with a direct government is probably a few thousand citizens (a city state). Giovanni Sartori (1967:254–256) is sceptical about the benefits of direct democracy or real self-government as the Greeks practised it in comparison with systems of representative government. He argues that a system based on direct personal participation demands that the citizens spend too much time and energy on the affairs of the state. The basic advantage of indirect systems of government is that the citizen can devote time to the necessary non-political activities of society but such systems also obtain an element of stability when political decision making goes through numerous intermediaries. Direct government that requires the actual presence and participation of the citizens is simply not feasible when vast territories and entire nations are involved.

James Madison argued long ago that oppressive factions could be better controlled in large states (republics) as various counterbalancing forces were more likely to exist in larger states:

> The smaller the society, the fewer probably will be the distinct parties and interests, the more frequently will a majority be found of the same party; and the smaller the number of individuals composing a majority, and the smaller the compass within which they are placed, the more easily will they concert and execute their plans of oppression. Extend the sphere and you take in a greater variety of parties and interests; you make it less probable that a majority of the whole will have a common motive to invade the rights of other citizens; or if such a common motive exists, it will be more difficult for all who feel it to discover their own strength and to act in unison with each other. (Madison, 1988:113–114)

Historical experience does not confirm this view of James Madison. Large states have been far from immune to the vice of powerful oppressive factions as the case of Germany (the rise of fascism) and several states in South America (military juntas) show.[4]

The possibility of face to face contact is an important factor that contributes to greater citizens effectiveness in small states. A small society is characterised by particularism rather than universalism in political actions. Dahl and Tufte (1974:96) hypothesise that: 'Within a country governed by a system of representative democracy, the larger the political unit the greater the extent to which public conflict is expressed and resolved through formal and impersonal organizations rather than through informal, face to face negotiations by the antagonists themselves'. They refer to two studies in support of their hypotheses, both based on local government studies in the USA and Sweden.

Such face to face societies have both benefits and costs. In a small state such as Iceland it is not difficult for the citizen to contact directly members of the parliament and there is a great likelihood that some friend or relative does personally know someone in a high office. The 'distance' from the citizen to the government is much shorter than in a larger society. Communication is more direct and the authorities should be more successful in solving various problems, because they are able to base their solution on a direct knowledge of the circumstances of their people.[5] On the other hand it is more difficult to govern by general principles and all kinds of personal involvement may lead to mistakes in government, especially mistakes in financial management. One example is the recent collapse of the financial system in the Faeroe Islands.[6]

The question is whether technical progress such as advances in communication can replace the function of face to face contacts between the representatives of government and the citizen. The form of communication, whether it is a computer on the Internet, a conversation on the telephone or a letter is not likely to affect the number of messages that a representative of the government can deal with personally or decrease the time spent on each message. Changes in the media are not likely to increase communication between leaders and citizens although it may be easier for some remote citizens to contact their representatives by modern means of communications. Modern communications have, however, doubtless made it easier to obtain knowledge of various social situations. The huge volume of information that flows in modern societies is mostly processed as statistics that is valuable for various institutions and helps in the decision making of the government. Technical progress has not changed the conclusion that citizen effectiveness increases as the state gets smaller.

Determinants of system capacity

System capacity is mainly determined by the productive capacity of the economic system. Economic growth of small states is largely dependent on the growth of foreign exports as the home market is too limited to absorb the output from efficient industries. Success in foreign trade is obtained by investment in human capital, prudent exploitation of natural resources and progressive government.

Economic problems especially in small developing states often make the state unable to offer suitable employment for the available labour force. Small states such as Cape Verde and sometimes Iceland in times of depression have exported employment problems to other states (larger states such as Turkey also solve their employment problems by emigrating workers). Remittances from emigrant workers are an important part of the foreign earnings in several small states (Edward Dommen and Philippe Hein, 1985:169–170). Special health services and higher education have also been obtained from larger states when it is too costly for small states to provide such services. This is another example of the effects of indivisibilities. Small states also lack the capacity to handle important problems, such as defence, that are both economic and political in nature.

The capacity of the state to fulfil the wants of the citizens also depends upon the level of autonomy or real independence of the state. An independent state has more freedom to manoeuvre and respond to changes than a state that is dependent upon other states for vital political or economic functions. Independence increases, however, the load upon the political and economic system and suboptimality in terms of military power or the 'weakness' of the state becomes more pronounced as the state becomes smaller and more independent. Real independence as a measure of system capacity becomes an indirect criterion of optimality.

Constraints on optimality

A state contains several subsystems of which the economic system and the political system are the most important. A state of an optimum size contains a combination of subsystems each of an optimum size (where an optimum often implies maximum efficiency). Several authors have advanced criteria for performance or efficiency that apply to various subsystems of the state. They can be viewed as constraints on optimality or, alternatively, as conditions that must be fulfilled to obtain the optimum size of the state. Conditions for an optimal size of the various subsystems will be discussed in this section.

Size and problem solving

The relation between the size of a state and its ability to solve problems is viewed as a problem of extending the boundary of the political system by Dahl and Tufte (1974:129–130, 134). A political system is too small for a problem if it lacks authority to secure compliance from actors (presumably outside the system) whose behaviour results in significant costs (or loss of benefits) to members of the system. Dahl and Tufte (1974:130–131) suggest the following strategies to solve the problems of small size:

1. The small system may adjust unilaterally to the behaviour of the outside actor or actors. It may simply accept the costs of that behaviour. [...] Alternatively, it may search for unilateral adjustments that will reduce or eliminate these costs. For example, if upriver towns pollute the water supply, it may install a purification system.

2. The small system may try to achieve a mutual adjustment by negotiation or bargaining. This process may be implicit or explicit.

3. By cooperating with other systems, a small system may succeed in helping to create a superior authority with boundaries large enough to include the relevant actors. This authority may be a confederation in which enforcement is carried out by the constituent parts, each acting within its own boundaries, a federal system with direct enforcement of federal laws and rules by a central government, but with some autonomy constitutionally reserved to the constituent units; or a unitary system, with no constitutionally autonomous constituent units.

They state that the above strategies are used by practically every state to meet the need for a variation in the size of political units whenever existing boundaries are too large or too small (Dahl and Tufte, 1974:135).

The political system may also be too large if the application of uniform rules throughout the system imposes costs (or loss of benefits) on actors that could be avoided by non-uniform rules. The alternatives for citizens adversely affected by uniform rules in a large system are to:

1) Adjust unilaterally, either by accepting the losses or by discovering an alternative that avoids them.

2) Engage in mutual adjustment by negotiation and bargaining.

3) Create a subordinate authority with boundaries small enough to include only the disadvantaged actors by means of administrative or legal autonomy within a unitary system (decentralization), a federal system with constitutional autonomy for the smaller constituent units (federalism), or a confederation (confederalism).

4) In the extreme case, separation into an independent sovereign system: the 'nation-state'. (Dahl and Tufte, 1974: 133–134)

Dahl and Tufte (1974:135) conclude that there is no optimum size for a political system, different problems require political units of different sizes.

Problems that cannot be handled even by large states are often the subject of global environmental policies and international law. If environmental problems require a global approach the larger states, in terms of population, may not have more influence than small independent states. This applies especially to the small states that control relatively large areas of the global territory. In international law size is not important at least if such laws are generally based on natural rights (whereas national laws could be based more on specific circumstances).[7] The principle of subsidiarity introduced in the Treaty of Maastricht by the European Union can be interpreted as a solution to the dual problem of capacity and disadvantages of uniform rules as will be discussed later in connection with regional integration.

Economic externalities and the scope of government

The optimum size of a state must fulfil the condition or constraint that the scope of the government is sufficient to control diseconomies, such as pollution, resulting from economic activities. Divergence between private and social costs results in inefficient use of resources. A government can deal with that problem through suitable taxation for example. Political activities may, in a parallel manner, lead to inefficient or unacceptable

25

allocation of values not intended or foreseen by the individual actors. The idea of using the theory of economic externalities to find the optimum size of government or the optimum city size resembles the idea of adjusting the boundaries of the political system to different scopes of problems.[8]

Gordon Tullock (1969:19–21) argues that the governmental unit chosen to deal with any given activity should be large enough to internalise all of the externalities which that activity generates. The size of the governmental unit depends partly upon the percentage of externalities that is required. Tullock shows that it is only this externality criterion that is relevant to the choice of governmental unit but not the possible economies of scale in providing services. The economies of scale are only relevant if a governmental unit itself must produce the particular governmental service. If the service can be purchased from a specialised producer, then the economies of scale cease to have relevance to the decision of the size of the governmental unit.

Tullock (1969:21–25) further argues that the optimal size of the governmental unit is normally smaller than is necessary to internalise all externalities. The first reason is that the smaller the governmental unit the closer it will fit the preference patterns of its citizens. Using an argument based on voting logic he shows that the smaller the governmental unit the less cost will the individual suffer from governmental activities of which he disapproves. The second reason is the declining degree of satisfaction as the unit of choice is raised. This is due to the increased difficulty of providing as varied range of choice if the unit of choice is large. Also the information content of the total communication between the people and the government is reduced by grouping decisions into large bundles. The third reason is that the cost of bureaucracy will probably increase faster than the size of the governmental unit.

Although it is not always clear when the 'size' of the governmental unit in this argument refers to the number of citizens or the scope of government, Tullock has shown that there are constraints on the size of government that diminish the benefits obtained by internalising externalities.

Purchases of governmental services from specialised foreign producers are hardly feasible in most small island states, although transport and communication costs have fallen rapidly in the recent years. Therefore the existence of economies of scale in the production of government services is relevant in small states. It does not seem, however, that small states are more costly as a governmental unit than larger states in many areas such as administration. In other areas it is clear that unit costs of public services rise rapidly if population falls below certain minimum levels. This is the case in health services, education and other areas where indivisibilities come into effect. In Iceland employment in central and local government was on the average 16.7% of total employment in the period 1980–1988. This is slightly higher than OECD with 15.2% but lower than the other

Nordic countries with 26.3% (Þjóðhagsstofnun [The National Economic Institute], 1991b:37). Some international services also are not within the means of small states. This applies to international representation such as the number and size of embassies abroad. A well organised foreign service and participation in international organisations can reduce the possible disadvantages resulting from smallness in this area.

External economies play a great part in problems related to investment and economic growth. An investment may be profitable for the economy as a whole, by creating external economies, although it may not be as profitable from a private point of view. Tibor Scitovsky (1969b:245–250) shows how an investment in one industry may lead to expansion and further investment in other industries. Such pecuniary external economies exist when producers are interdependent through the market mechanism. Efficient plant capacity does not vary between countries but a new plant is generally a much larger percentage of an industry's total capacity in a developing country. In developing countries investment is likely to have greater effects on prices and create greater pecuniary external economies. Divergence between private profit and social benefit is therefore greater in developing countries according to Scitovsky. This also applies to private and social costs.

The size of the state or the internal market determines the occurrence and magnitude of pecuniary external economies. In a small state there are few industries and the largest of these are usually export oriented. Interdependence between producers is therefore not significant in most cases. Greater divergence between private profits and social benefits in small countries requires greater scope of the government to internalise externalities resulting from investment. The ability to control or internalise such external economies is dependent upon the size of the state. Generally it is a matter of the scope of the government rather than the size of the state in terms of population that determines how the state deals with external (dis-) economies. Some diseconomies such as pollution may be easier to control in a small state. The existence of pecuniary external economies, or the divergence between private profits or costs and public benefits or costs, imposes a constraint on the optimum size of states.

Size and self-regulation

Efficiency in self-regulation and goal seeking are important constraints on the optimum size of states. Karl W. Deutsch (1967:283) argues that the efficiency of a feedback process can be evaluated in terms of the number and size of its mistakes, that is, the under- or over-corrections it makes in reaching some goal. He draws an analogy from a goal seeking system where the relationship between four quantitative factors determines the outcome of a goal seeking process. These are *load, lag, gain* and *lead*. Deutsch (1967:285–286) adds that a feedback model of this kind permits us to ask a number of significant questions about the performance of governments that are apt to receive less attention in terms of traditional analysis:

1. What are the amount and rate of change in the international or domestic situation with which the government must cope? In other words, what is the *load* upon the political decision system of the state? [...]

2. What is the *lag* in the response of a government or party to a new emergency or challenge? How much time do policy makers require to become aware of a new situation, and how much additional time do they need to arrive at a decision? What is the lag in the response to new information that is brought into the political decision system through one channel rather than another - for example, the lag in the reaction to information that is reported more or less 'straight to the top' in contrast to the information that is first accepted among some particular social or occupational groups? [...]

3. What is the *gain* of the response—that is, the speed and size of the reaction of a political system to new data it has accepted? How quickly do bureaucracies, interest groups, political organizations, and citizens respond with major recommitments of their resources? [...]

4. What is the amount of *lead* that is, of the capability of a government to predict and to anticipate new problems effectively? [...]

Deutsch (1967:286) further argues that this approach is useful in evaluating political systems. 'If we assume, [...], that all governments are trying to maintain some control over their own behaviour, to maintain as long as possible the conditions for the existence of their political systems, and to get nearer to, rather than further away from the goals that they have accepted, then it would be possible to evaluate different configurations of political institutions in terms of their capacity to function as a more or less efficient steering system.'

The first impression is that small states do not seem to be suboptimal compared to larger states when the capacity of their political system is evaluated in terms of the Deutsch criteria. A small state is no less affected by changes in the international system than larger states but domestic demands produce similar load. Information reaches the top level faster in small states and the lag in information processing is therefore less in internal matters. This does not apply to foreign policy where larger states have more capacity than small states. Only in some situations may adjustment be faster in small states although mobilisation of resources is probably easier. In Iceland the government has generally been less progressive in many areas of the economy, civil rights and public services than the neighbouring countries, although eventually legislation in Iceland has followed the Scandinavian countries in most areas. Trade, for example, first became reasonably free in 1960 but most tariffs stayed very high until Iceland joined EFTA in 1970. The capability of a government to predict and to anticipate new problems effectively is probably not better in a small state as the government in a large state may have better access to specialised research or secret information.

When actions of states are viewed in terms of Deutsch's criteria as a self-regulating goal seeking process it does not seem that a small size of a state is a serious disadvantage. The role of information will be further discussed in the next section.

Size and information processing

The political system of a state has been viewed as an information system with input, output and feedback channels. The capacity to make authoritative allocations of values can also be subject to efficiency criteria. Greater efficiency means that a larger volume of demands could be processed to the satisfaction of the population. The government or other institutions of the state must obtain information on the preferences or demands of the citizens. Information is not free, however, and the capacity of the state to process and collect information is subject to several constraints.

Kenneth. J. Arrow (1974:38–39) discusses four key characteristics of information costs or the inputs needed for the installation and operation of information channels. First and most important is the limited capability of the individual to acquire and use information. Although use of computers and technical progress may expand the capacity this limitation is a fixed factor in information processing and one may expect a sort of diminishing returns to increases in other information resources. The second characteristic is that information costs represent an irreversible investment that arises because of the need to spend time and effort to be able to distinguish one signal from another. The third characteristic is that information costs are not uniform in both directions. Common background and former experience make it easier to process information or open information channels between individuals. Arrow (1974:53–55) adds that the functional role of organisations is to take advantage of the superior productivity of joint actions. The economies of information in an organisation occur because

29

much of the information received is irrelevant and it is the reduction in re-transmission which explains the utility of an organisation for handling information.

When the number of members of a society rises the information flow increases and as the capacity to handle the necessary flows of information is always limited, waiting time increases faster than proportionally. A single government of a nation of 200 million people cannot know the needs of their countrymen as well as a government of 200,000 people. In practice this problem is solved by dividing the responsibility of authoritative allocations between different stages of government, local and national or local, state and federal government. This division shows that from a pragmatic point of view there is an upper limit on the size of government in each specific area. In large firms special methods of management are needed to cope with increased size and complexity. From this it can be deduced that there is a tendency for efficiency in information processing to diminish with the size of the political system where the size refers to the number of members approximately. Therefore the functions of the state that depend on processing of information from the citizens and responses to the needs of the individual are less efficient as the size of the state increases. The capacity to process information in conjunction with other aspects of citizens effectiveness is the main constraint that prevents the optimum size of states from being a function of system capacity only.

Market factors

The free market, in theory, reflects the preferences of the consumers within their budget constraints. The consumers influence the allocation of values through the market mechanism. It does not seem that this function of the market or that how demand and supply for goods and services determines prices is limited in any way by a small size of a state. On the other hand a small state is more constrained in its productive capacity.

Scitovsky (1960:283–284) suggests two criteria of optimum economic size. He calls an economy technologically too small if its market is too small to provide an adequate outlet for the full capacity output of the most efficient productive plant in a given industry. From this it follows, he says, that the minimum size of an economy is generally different for different industries. An economy that is large enough to provide an adequate market for at least one optimum sized plant in all industries producing final products may still be suboptimal if some of those plants provide a market that is too small for efficient production of needed intermediate products. An economy is economically too small if it fails to provide the competitive conditions necessary to spur to utmost efficiency and to lead to the establishment of the technologically most efficient plants. Scitovsky adds that small states can hardly rely on export markets to overcome these disadvantages mainly because export markets are unstable and can be closed because of balance of payments difficulties or for political reasons. He concludes that if an economy is too small technologically, economic union is better than international trade if it guarantees free and unrestricted trade as well as complete stability of exchange rates among members of the union.

The economic system of small states in autarky is doubtless suboptimal both economically and technologically but the world market is more stable now than Scitovsky argues. It is also not clear that economic union is as beneficial for small states as he argues. World trade has steadily expanded and international trade has been made freer and safer through GATT and other international trade regimes. There seems to be little danger of export markets being closed except in times of severe conflict. Prices will always be unstable but the negative effect of unstable prices can be reduced by suitable policies such as price-equalising funds. International competition can no longer be avoided by entering an economic union. Most states are now members of the GATT (or its successor the World Trade Organisation) or apply its rules on a *de facto* basis. Therefore there is less scope for using tariff walls and other trade restrictions to protect industries from competition and world market prices. Technical progress, especially in communications, has also opened up new possibilities and reduced the effect of a small home market. Not all goods and services are traded internationally. In the case of small island states, long distances make it uneconomical to trade in bulky goods or services that require close contact between producers and consumers. In these areas the smallness of the economy cannot be escaped by foreign trade. The autarky situation is, however, of limited importance in practice compared to the importance of open markets and foreign trade.

Foreign trade of small states and economic integration will be discussed in detail later but it is clear that foreign trade increases the capacity of the economic system of small states and reduces the negative effects of market factors.

Size and institutional framework

Mikael M. Karlsson (1992) discusses ideas based on biological observations that imply that size and form are closely related in social structures. A thing of a certain form must be of a certain size. Every human society must adopt a form of institutions and a political and economic system that suits its size. A small state must use an institutional framework that fits its size and needs. He criticises Iceland for having imported economic and political institutions from much larger states without adapting their structure to the

31

size and special needs of Icelandic society.

This idea of Karlsson suggests a new constraint on optimality, namely that the form and structure of (public) institutions and organisations should fit the size of the state that they operate within. An institution that is not of a suitable form for a small state will not be efficient and therefore increase the suboptimal character of the state. Just as uniform rules of a too large political unit impose cost on actors there is a cost resulting from unsuitable size of institutions. An example of an institution that has been 'imported' from larger states is the Central Bank of Iceland. Iceland established the Central Bank as an institution in 1961. Until that time the issue of money and other functions of money management were entrusted to a small division of the National Bank of Iceland, a state owned commercial bank. Monetary policy will be discussed in a later chapter but it is clear that the existence of a central bank did not prevent extremely high inflation in Iceland in the seventies and eighties. There is no doubt that monetary management could be greatly simplified by pegging the currency to ECU, for example, and by limiting monetary growth to the growth of the national product. Such a policy would not require a central bank in the shape of a large institution as is the case in Iceland today. Some other functions of the central bank such as the management of international reserves and supervising of the financial market can be done in other institutions or ministries.[9]

Spatial efficiency

To the constraints above may be added a special criteria or constraint that refers to efficient spatial organisation of the state. The various social systems of the state may be abstracted from the geographical dimension but the population is located in certain places. The distribution of the population affects economic growth and political strength. Centre-peripheral forces or polarisation effects that arise mainly because of unequal economic growth are also important determinants of spatial organisation.

Difficult terrain or long distances between settlements may hinder growth of cities which is a prerequisite for the development of many services and industries. In larger countries cities may be too large or the settlement structure otherwise cause pollution and damage to the environment. If the territory of the state contains many isolated small settlements, indivisibilities become more pronounced in many activities such as health services and education. It is also more difficult to develop new industries to augment land-based activities and create jobs for a growing population. A dispersed and fragmented settlement structure leads to extra communication costs and communication within the society is diminished. Inefficient spatial organisation or a regional problem will make the state effectively smaller and weaker and is therefore a constraint on the optimum size of states. Spatial efficiency and other aspects of spatial organisation relating to the area factor will be discussed in the next chapter. Centre-peripheral relations are the subject of chapter 8.

Summary

The concept of the optimum size of states is an important but a complicated concept. The main criteria for optimality are system capacity and citizen effectiveness. Other constraints and criteria that apply to subsystems or important functions of the state include capacity to solve problems, the scope of the government, the ability of the state to seek goals and self-regulate its actions, efficiency in processing information, institutional framework, spatial organisation and market factors. The theory of the second best shows that a small state composed of suboptimal systems may obtain a satisfactory overall status in a second best world. The first impression is that the citizens in small states have at least as much influence on the actions of the state as citizens of large states but in areas related to political power and economic capacity small states are clearly vulnerable. In the following chapters the main political and economical factors that influence system capacity or the effective size of a state will be studied, starting with the area factor.

Notes

1   See Albert Einstein (1954:118–132) on this idea of a world government.
2   In economics an allocation is Pareto efficient if no reallocation of commodities between consumers or factors of production between producers can make someone better off without making someone else worse off. In a perfect world, it is assumed that policies to correct inefficient allocation of resources will not affect policies of income distribution. Allocative efficiency and distributional issues can therefore be separated out. See Peter Jackson (1992:105–114).
3   See Dahl and Tufte (1974:4–12) for a short survey.
4   Totalitarian states such as the former Soviet Union and China can be included although they have no experience of democracy.
5   Population per lower house member is generally much lower for small states than for larger states, see table in Denis J. Derbyshire and Ian Derbyshire (1991:99–101).
6   The historical background and the events leading to the collapse of

33

the financial system in the Faeroe Islands are described by Eðvarð T. Jónsson (1994).

7   See the discussion on the legal concept of independence in chapter 4.

8   See George Tolley and John Crihfield (1987:1289–1294).

9   For a discussion of the role of central bank in small states see Deena R. Khatkhate and Brock K. Short (1980).

3   The area factor

In the last chapter the state was mostly regarded as a social system without a geographical dimension. Only the criteria of spatial efficiency related the size of the state to spatial organisation. Geographical conditions have important but conflicting consequences for the functions of the state especially in combination with a small population. Two aspects of the area factor will be discussed here. The first is the increase in area which occurs when a state declares a 200 mile Exclusive Economic Zone (EEZ). The second is the scattered settlement structure which characterises many small island states. It turns out that the effective size of a state is enlarged if the surrounding sea or the Exclusive Economic Zone is counted as an integral part of the territory. The boundary of the political system is extended and the capacity of the state is increased. Global environmental problems such as marine pollution may, however, increase the load on the political system. Regional factors leading to isolated or scattered settlement structure over a large sea or land area make the state effectively smaller and also increase the load on the political system. Examination of the case of Iceland confirms this conclusion. Technological advances in communication coupled with suitable regional policies help to reduce the negative contributions and strengthen the positive contributions of the area factor to the optimum size of states.

The role of the Exclusive Economic Zone

The effective geographical size of many states especially small island states increased greatly when a 200 mile economic zone was adopted. Control of large areas of the sea undoubtedly adds to the resource base of small states

35

and increases their sphere of influence but at the same time their responsibility to manage and protect this resource is increased.

Development of the EEZ

Although the doctrine of the continental shelf was revived in Latin America in the second decade of this century the era of extensive maritime claims after World War II opened with the Truman Proclamation of 1945, where the United States claimed the natural resources of the subsoil and sea-bed of the continental shelf to be subject to its jurisdiction and control. Also proposed was the establishment of fishery-conservation zones in waters contiguous to its coast but beyond the 3 nautical-mile limit that was then the generally accepted limit of the territorial sea. In the following years several South American states unilaterally declared extensive maritime zones. Their claim went beyond the Truman Proclamation in that a single comprehensive claim of sovereignty was made over the shelf and its superjacent waters.

In 1958 the first United Nations Conference on the Law of the Sea (UNCLOS I) was held in Geneva. There four draft conventions were discussed relating to (i) the territorial sea (ii) the high seas (iii) the continental shelf, and (iv) the conservation of fisheries. No agreement was reached on the extent of the territorial sea or of fisheries limits although twelve miles became the generally accepted limit for the territorial sea as a consequence of the Territorial Sea Convention in 1958. In the second conference in Geneva in 1960 the issue of territorial limits was not resolved. In 1973 the third conference (UNCLOS III) started and after a series of sessions concluded in 1982 with the adoption of the UN Convention on the Law of the Sea.

Prior to this conference the subject of exclusive economic zones had been the subject of discussion at the regional level by Latin American and African states. In the Declaration of Santo Domingo for example it was stated that a coastal state had 'Sovereign rights over the renewable and non-renewable natural resources ... of an area adjacent to the territorial sea called the patrimonial sea' (Mpazi A. Sinjela, 1989:63). The objective of coastal developing states in establishing an EEZ was to protect their resources, especially fishing resources, from the exploitation of other states. This was also the main idea behind the policy of Iceland in its fight for a 200 mile limit which resulted in the 'cod wars'.[1]

Within the Exclusive Economic Zone, which extends beyond the territorial sea, all states continue to enjoy *jus communicationis*, but the coastal state exercises sovereign rights with regard to all natural resources and other powers deriving from, or complementary to those rights (Barbara Kwiatkowska, 1989:XX). The EEZ may be defined as follows:

The exclusive economic zone is an area beyond and adjacent to the territorial sea [TS] that extends up to 200 miles from the TS baselines, in which the coastal state has sovereign rights with regard to all natural resources and other activities for economic exploitation and exploration, as well as jurisdiction with regard to artificial islands, scientific research and the marine environment protection, and other rights and duties provided for in the LOS [Law Of the Sea] Convention. All states enjoy in the EEZ navigational and other communications freedoms, and the land-locked and other geographically disadvantaged states [...]—specific rights of participation in fisheries and marine scientific research. (Kwiatkowska, 1989:4)

The EEZ is however not a part of a sovereign state in the same sense as its territorial waters. David Attard (1989:309) says that today neither sovereignty nor freedom provides an acceptable basis for a viable regime to regulate the use of the sea beyond the territorial sea. 'The EEZ proposed by the 1982 Convention is a contemporary approach to this dilemma which attempts to constrain both sovereignty and freedom in the common interests; it is neither an extension of the territorial sea nor part of the high seas.'

Small states and the EEZ

The Exclusive Economic Zone is an important concept for many small states. Many are islands and can claim an extensive economic zone in accordance with The Law of the Sea Convention. In 1988, 23 small states had established an Exclusive Economic Zone of 200 miles or a Fisheries Zone of 200 miles. Not all small states will, however, benefit from this new regime. Six states (Luxembourg, Swaziland and four 'minute' states in Europe) with a population of under one million belong to the class of land-locked states. Although under the 1982 Convention on the Law of the Sea landlocked states are allowed to participate in the EEZ in the same region or subregion to which they belong, the provisions of the law are complex and are likely to lead to disputes in their actual determination, allocation and eventual exercise (Sinjela, 1989:73). During UNCLOS III several small states associated with a group of states known as Geographically Disadvantaged States. These states were perceived as having limited access to the sea and its resources. Several small states in the Caribbean have not gained from the LOS convention because they lost their fisheries' rights to coastal states like Venezuela. In the case of the Eastern Caribbean, small states obtained EEZ to 'some of the most biologically unproductive waters in the region' (Dolman, 1985:56). Several developing small island countries do not yet have the capacity to exploit the EEZ or enforce their rights there despite the potential existence of considerable marine resources. Dolman

(1985:61) states that the advantages of a large EEZ are more theoretical than real while the small states in the Pacific and Caribbean are highly dependent on others to exploit their marine resources.

Control of the fishing banks around Iceland through the extension of its fishery limits has been a central objective of Icelandic foreign policy. When a 50 mile fisheries limit and subsequently 200 mile EEZ was adopted, foreign vessels lost their share of the catch around Iceland. According to Sigfús Jónsson (1981:4) more than half of the roughly 700,000 metric tonnes of demersal fish that were harvested in Icelandic waters in the early 1930s was taken by foreign vessels. The total catch of demersal fish as well as the share of foreign vessels diminished gradually in the following decades. After the adoption of a 200 mile EEZ the total catch has usually been between 500 and 600 thousand metric tonnes (in cod equivalent units). The adaptation of an EEZ led to an expansion of the fishing industry in Iceland and contributed largely to the rapid growth of the economy in the seventies as will be discussed later. The existence of the EEZ has also affected Iceland's foreign policy in matters of regional integration. National control over the EEZ is a fundamental policy in foreign relations. This position is, however, not compatible with the present common fisheries policy of the European Union. Full membership of the European Union is unlikely unless Iceland retains full sovereignty over the EEZ.

Although the small islands that do obtain large Exclusive Economic Zones may not as yet have the capacity to exploit their maritime resources, the adoption of the EEZ significantly changes their status. The rights of a state over its EEZ, although not equivalent to territorial rights in international law, are so extensive that the EEZ may be considered a part of the state that has sovereignty over it. This means that the territorially small states are not so small any longer and their importance in international relations greatly increases.[2]

Small states and the global environment

Rapid population growth and environmental decay is of increased concern in the global system. Although many small island states are remotely placed in the Pacific or Atlantic ocean they are not immune to the dangers of environmental decay. Besides the special measures needed to protect vulnerable biological systems of small islands small states must also support international regimes that seek to control global environmental decay, especially in the oceans.[3]

The report of the Club of Rome in 1972 demonstrated that there are limits to economic growth in a closed system if it is based only on increased volume of production without better utilisation of raw materials or capital.[4] The report stated that 'If the present growth trends in world population, industrialisation, pollution, food production, and resource depletion continue unchanged, the limits to growth on this planet will be reached sometime within the next one hundred years. The most probable result will be a rather sudden and uncontrollable decline in both population and industrial capacity' (Meadows, 1972:23). The main criticism that can be directed towards the Club of Rome report is that the flexibility of the market economy in allocating scarce resources through the price system is underestimated. Generally it seems that the assumption of a more or less fixed resource base for an economy is of a doubtful value in predicting the growth potential of an economy. Economic growth can be driven by the increase in production and trade of service industries that traditionally require limited input of raw materials.

Although some of the small states in our class are relatively densely populated they are not in most cases concerned with the negative aspects of population growth. Lack of resources and population pressure has, however, been considered a problem for small states. Both R. G. Ward (1967) in discussing population trends in Polynesia and D. A. G. Waddell (1967) in discussing problems of British Honduras (Belize) were concerned with problems of overpopulation relative to the narrow resource base in these areas and were sceptical about the ability of these areas to support increased population. Since this was written the population of Belize has doubled. Economic growth has also been 5% per annum in the last decade in those states. Several states in Polynesia have experienced similar trends. Population pressure is not a problem in small states, on the contrary, their population is probably too small in many respects.

New environmental problems have been discovered since the report of the Club of Rome was published, for example the depletion of the ozone layer and increased 'greenhouse effect' due to higher carbon dioxide levels in the atmosphere. Rising sea levels are especially of concern to low lying island states.[5] Many of the major destructive forces leading to a collapse in the models of the Club of Rome are therefore still operative.[6]

> Already the planet's degradation is damaging human health, slowing the growth in world food production, and reversing economic progress in dozens of countries [...]. The decline in living conditions that was once predicted by some ecologists from the combination of continuing rapid population growth, spreading environmental

degradation, and rising external debt has become a reality for one sixth of humanity. (Lester R. Brown et al., 1992:175–176)

If the worst scenario of the environmentalists becomes reality, with widespread famine and collapse of the ecological system in large areas, international relations will suffer severe strain. In such circumstances it would be tempting to violate the rights of small states if they still have available unpolluted land, water and food. It is, however, not too late to adopt a policy that helps to prevent an ecological collapse of the global system.

Environmental policies of small states

The conservation and protection of the sea is probably the most important issue for small island states or archipelagos that control large areas of the surface of the earth within their 200 mile economic zone. Many small states, especially island states, have a very fragile ecological system which is easily disrupted if new species are introduced. Ecological disasters such as oil-tanker accidents are a possible threat where small states are close to main oil-tanker routes as pointed out by Dommen (1985:42). Iceland and other small states have, due to the lack of large cities and manufacturing industries, been able to avoid severe problems of industrial pollution and waste although they are affected by pollution created by other states.

Through the institution of international regimes the small states have the possibility to become active in global environmental politics. An international regime can be defined as 'a system of norms and rules that are specified by a multilateral legal instrument among states to regulate national actions on a given issue' (Gareth Porter and Janet W. Brown, 1991:20). Global environmental regimes have been established to protect whales, limit trade in endangered wildlife species and hazardous wastes, limit long-range transboundary air pollution, protect the ozone layer and limit marine pollution from ships and the dumping of wastes and other materials in the oceans (Porter and Brown, 1991:21). Such regimes based on multilateral legal instruments allow small states to act formally on an equal footing with larger states. The global environmental policy of small states should be to preserve the environment with strict controls on pollution. A policy of preservation should, however, not hinder prudent utilisation of the available resources within the 200 mile economic zone of each state.

Regional problems and spatial organisation

When the area factor generates severe regional problems the effective size of a state becomes smaller in a political and economic sense. It seems that many small states face regional problems of much larger dimensions than larger states where the peripheral regions are often a much smaller fraction of the country. In small island states, especially archipelagos, the internal market is often fragmented and the provision of social services and general administration becomes more difficult because unit costs of providing infrastructure increase fast as the settlement structure becomes more scattered. Problems may be intensified if geographical division into regions coincides with cultural and ethnic divisions. In small island states the regional problem is not only a problem of redistribution but rather an integral part of the problem of economic development and economic viability.

It is possible to distinguish between three cases that are relevant for spatial organisation and spatial efficiency by considering the topography of small states. The cases are polarised island states, landlocked states and archipelagos or large sparsely populated islands.

First, there is the case of moderately scattered island states such as the ones found in the Caribbean, where the state is polarised, politically and culturally, between two (or more) main islands often at different stages of economic development. This situation has created tensions and separatist tendencies in St Kittis and Nevis, Barbuda and Antigua and Trinidad and Tobago. Anguilla had already left the St. Kitts-Nevis-Anguilla association and Nevis announced its intention to leave the St. Kitts-Nevis federation in 1992.[7] François Doumenge (1985:102–103) states that political fragmentation is normal with islanders unless under the pressure of external forces. In archipelagos the tendency is towards total fragmentation or frequent antagonism between the 'windward' and 'leeward' communities in the case of the largest islands. Such fragmentation runs counter to demographic, social and economic needs because the society must be above a certain threshold of viability of various kinds, such as natural resources, population, production activities and cultural relations if it is to function harmoniously.

Second, there are landlocked continental small states such as Luxembourg. Such states do not face an internal regional problem. They are often able to use the infrastructure of their larger neighbours and their economies are integrated with the economies of the surrounding regions.[8] This possibility is likely to increase somewhat their system capacity and effective size.

Third, there is the case of archipelagos and large sparsely populated countries. In the Pacific and Indian Ocean archipelagos are often spread over vast areas of ocean space with scattered settlements isolated from each other as well as from other countries. For instance the Maldives, population 223,000, contain close to 1,100 islands in 19 atolls spanning an arch of over 800 km and around 200 of them are inhabited.[9] In Greenland it is not the sea that separates the settlements but vast areas of uninhabitable land. In these cases administration and the provision of infrastructure are expensive and transport and communication costs are relatively high. Efficient spatial organisation of small states scattered over many islands or relatively large in area will be discussed below.

Regional problems in sparsely populated territories

The population of small states is often dispersed in settlements over a number of small islands, in the case of archipelagos, or in remote fjords and valleys, in the case of Iceland and Greenland. Many rural areas of larger nations are also sparsely populated. Scattered settlement structure and low density of population leads to special problems related to distribution of public goods that affect the spatial efficiency of small states.

The economic problem is the high cost of providing modern services and basic infrastructure in sparsely populated places. Unit costs of provision of education and health services as well as indivisible basic structures such as roads and harbours are generally much higher in sparsely populated regions than in densely populated regions. If these services are not available people are likely to migrate to the places where such services can be found.[10] In Iceland as in many other states it is the capital area that grows (figure 3.1). Cities are needed for the development of many economic and cultural activities but natural resources will be under-utilised and the city congested if a large part of the population migrates to the capital. Depopulation of the countryside means that particular ways of living and cultural values are sacrificed for the mass culture of the cities. Also lost are technical skills adapted to specific circumstances and developed by many generations.

The political problem is whether people living in remote places have a justifiable right to demand the same level of services as people living in densely populated areas where such services are generally much cheaper and easier to supply. If this is the case the load upon the political system will be greatly increased. There is in this connection an important difference between sparsely populated developed regions (or small states) and densely populated developing regions. Investment in new roads for example in a sparsely populated region will generally always be under-utilised but in a densely populated developing region such an investment will become fully utilised as the economy develops. In a small state the problem of keeping the balance in the settlement structure of the country and providing all the inhabitants with equal basic services becomes an important political problem that can reduce the internal strength of the state. This problem will be discussed below.

Distributive justice in sparsely populated regions[11]

It is generally accepted that economic gain or profit seeking is an important stimulus for economic progress. Therefore it can be argued that inequality in income distribution creates greater benefits for the society as a whole than forced equality of incomes. In principle, the benefits of inequality can be shared by redistribution that makes everyone better off. The resources of the earth are also unevenly distributed and cities are required for the growth of many services. Unequal distribution of economic growth in geographical space (horizontal inequality) seems to be beneficial in the same way as unequal wages or income between individuals (vertical inequality) stimulates economic activity to a certain extent. Great difference in development between regions or countries is, however, not beneficial for economic growth or international trade.

It is necessary to distinguish between 1) inequality in development between countries and 2) geographical inequality within the same society or nation. Within each society it is the relative position of the individual that is of prime concern but not the geographical position of the individual (who has often the opportunity to migrate to the faster growing regions). Policies to reduce geographical inequalities can be unjust because inequalities between individuals or vertical inequalities are generally much larger than inequalities between regions or horizontal inequality. The worst off individuals can be living in regions where economic growth and incomes are the highest. This implies that inhabitants of slow growing regions do not have the right to claim compensation from inhabitants of fast growing regions on the basis of geographical inequality alone as an addition to the basic rights they may have as less favoured individuals in the society.

Distribution policies in small states

The state must try to provide the basic or primary goods and services, which include education, transport, electricity and water supplies and health services to every citizen. There are, however, certainly limits to the amount of basic goods that society can afford to provide in remote places if the unit cost (both construction and operation) of such services rises rapidly with increased distance from the centre. The question is how much one can tax the densely populated region (or city) to invest in remote regions without causing unacceptable loss of welfare to the society as a whole.

Investment in a public utility can increase welfare if total consumer

43

surplus is higher than the cost of the project.[12] In sparsely populated regions one cannot expect many such investments to be profitable (in this sense) even though users would be prepared to pay relatively more for access to such utilities. This means that from a welfare point of view many projects in sparsely populated regions are uneconomical if each project is evaluated separately. In the case of a distribution system such as broadcasting or electric network little or no loss of welfare results from the expansion of the system to some marginal places if the cost is spread over the system as a whole. For example only one radio transmission unit could cover a whole city. The price of using this service would be very low per user in the city and a large consumer surplus could be involved. This consumer surplus could in principle be used to justify additional investment in several radio transmitters that served remote places. The hypothetical limit of taxation allowed in densely populated regions would then be equivalent to the amount of consumer surplus that arises because of high density or the concentration effect in cities. But in a small country the taxing base might never be large enough to provide funds for adequate distribution of basic goods to remote places. One can expect regional problems to increase in developing small states, especially archipelagos, as their economies become more market and service oriented.

Geographical inequalities that the small state is unable to reduce weaken the small state internally. It would for example be difficult to denounce secession movements in remote parts of an archipelago if the inhabitants felt deprived of necessary services. Reducing regional inequalities and keeping a balance in the settlement structures will be an important task for many small states in the near future.

Regional problems and policy in Iceland

In Iceland more than half of the population, which is ethnically homogeneous, lives in the capital, Reykjavík, or the surrounding area. The rest of the population lives in small towns and villages or farms around the country. The villages are mostly dependent on fishing and the fish industry or services to agricultural regions. Dispersed settlements and harsh climate make the provision of infrastructure expensive and cost of services high. On the other hand Icelanders have been able to utilise local resources such as local fishing grounds, geothermal heat and pasture land.

Growth of the population in the last decade or so has taken place mostly in Reykjavík and the surrounding area as migration from the outer regions has been on the order of 1% of the population per annum for a long time. Many villages in remote fjords and agricultural regions have been experiencing depopulation which is due to declining agriculture and stagnation in the fishing sector but the attraction of city life and the more diversified labour market in the capital area also contributes to this trend. In figure 3.1 migration to the capital area is shown.

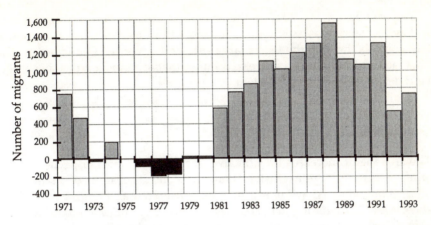

Figure 3.1 Internal migration in Iceland, 1971–1993. Net migration to the capital area

Source: Byggðastofnun [Institute of Regional Development in Iceland].

The viability of specific regions outside the capital area and the viability of the economy as a whole are closely connected because a substantial part of Iceland's export earnings originates in the dispersed fishing villages around Iceland. Agriculture provides less than 5% of total Gross Domestic Product but is an important part of the settlement structure of the country. Regional policy has therefore been closely connected to general economic policy, agricultural policy and fisheries policy.

Regional policy has been implemented through financial support to non-profitable firms, investment in infrastructure in remote areas, high subsidies to farmers and import restrictions on agricultural products.[13] Despite the high cost there has most of the time been widespread support for this regional policy.[14] It is clear, however, that changes in this policy are needed. In the fishing industry diminishing quotas, increased importance of auction markets and technical development seem to favour larger firms and concentration in fishing and fish production. This development makes some small fishing villages uncompetitive. In agriculture the GATT agreement will lead to a gradual reduction in import restrictions. Without import restrictions a large part of traditional sheep and milk farming is unprofitable. The most remote agricultural regions have few alternative development possibilities although tourism and modern communications open up some new possibilities. Increased unemployment and a general stagnation in the economy since 1988 makes the financial burden of the regional policy, especially the agricultural policy, more visible. It is therefore likely that regional policy in Iceland will be a source of increased political conflict unless more effective solutions to the problems will be implemented.

The area factor has shaped the economic, cultural and political life of Iceland. Iceland possesses abundant energy resources, rich fishing grounds and controls through the EEZ a large slice of the North Atlantic ocean. The area factor has effectively increased rather than diminished the size of Iceland despite the regional problem.

Conclusion

The area factor increases the effective size of small states, especially those which control a large Exclusive Economic Zone. The rights of a state over its EEZ are so extensive that it may be considered an integral part of the state. An EEZ therefore increases political power. An EEZ containing valuable natural resources such as fishing stocks is also a basis for economic growth as the case of Iceland shows. Environmental decay resulting from rapid population growth and pollution can reduce the benefits of an EEZ and requires active participation of small states in international regimes that are concerned with the global environment.

Regional problems arise easily when the settlement structure is scattered as is the case in many small island states. Advances in communication and transport have, however, reduced the disadvantages of remote location and made it easier to implement suitable regional policies. Regional problems should therefore not make the small state more vulnerable or weaker. In the next chapter the concept of real independence and the question of political vulnerability of small states will be discussed.

Notes

1   This brief historical account is based on Attard (1987:1–11) and Sinjela (1989:63–65).

2   It is likely that the resources of the sea will be of increasing value in the future, especially for food production.

3   For a discussion on biological 'endemism' in small islands see Doumenge (1985:76–78).

4   Economic growth in Eastern Europe has been based solely on volume increase in the input of raw materials, labour and capital. Productivity growth was not a driving force nor did efficient service industries develop sufficiently. When available productive factors were fully utilised at the given level of technology, economic growth

stopped leaving behind great environmental problems.

5   See for example Paul Kennedy (1993:104–121).
6   See Donella H. Meadows et al. (1992:xii–xvi).
7   According to Mark S. Hoffman (1992:794).
8   For instance Liechtenstein uses Switzerland's telephone and postal services (Dolman, 1985:43).
9   Dommen (1980:937) found that the average distance between islanders was 4 meters compared to 32 centimetres between continentals. He measured the distance between extreme points of each country in his sample divided by the number of inhabitants.
10   In a study of the demographic transition in islands John G. Cleland and Susheela Singh (1980) observed unbalanced growth in several Pacific islands where the capital or one island in an archipelago grew faster than the rest of the country. This was the case in Fiji, Samoa and Cook Islands for example. Urbanisation is, however, not as advanced in many pacific states as in Iceland. See Cleland and Singh (1980:980–985).
11   This section is based on Ólafsson (1991).
12   See Alexander M. Henderson (1969:541–560).
13   Government support of agriculture was 10.7 billion Ikr in 1992 but recent abolition of export subsidies has reduced this amount to approx. 7.7 billion Ikr or 2% of GDP in 1993 (OECD, 1994:49). To estimate the welfare effect of this policy the cost to consumers brought about by import restrictions and exorbitant prices of some domestic products must be added to this sum.
14   The Social Democratic party is the only party which has consistently opposed the present agricultural policy in Iceland.

4   Political power and independence

An independent state is here regarded as an organisation who has two main functions. First to manage social, economic and political affairs within its territory, for instance through legislation. Secondly, to conduct relations with other states.[1] Independence is not real if a state is highly dependent on other states for security or financial support or lacks the basic apparatus to supply its citizens with functions or services normally carried out by governments in other states. Small states have no significant armed forces and lack political and economic power. Small states are therefore suboptimal from the point of view of security and face significant problems of existing in an unstable world of larger powers. This state of affairs leads to the question whether small states are independent in the same sense as larger states and whether their weakness prevents them from playing some part in the international community. The answer obtained by using the theory of weak states and the dimensional approach to independence is that small states can increase their level of security and obtain a satisfactory position on the independence scale. The case of Iceland confirms this view.

The legal concept of independence

To be an independent or sovereign state in the legal sense its people must be living together in a defined territory under an organised government not subordinate to any other government.[2] The status of independence is formally obtained by *de jura* recognition of other states. The independent state is a fundamental unit in international relations and political analysis in general. The state is, however, not the only possible participant in the international system. Non-state actors such as multinational firms and liberation movements also participate in global affairs, as do organisations based on states, such as alliances, economic unions and the UN system. Non-state actors influence and modify the actions of states. Semi-independent states and regions are also important participants in international relations and their legal status is related to the legal status of independent states. The legal concept of independence or sovereignty is therefore more complicated than appears at first sight. This concept has also been changing and some of the changes seem to affect the status of small states.

Hans J. Morgenthau (1968:302–303) discusses different aspects of the concept of sovereignty. Three terms equivalent to sovereignty are discussed; independence, equality and unanimity.[3] Independence is a synonym of sovereignty that signifies the particular aspect of the supreme authority of the individual nation that consists in the exclusion of the authority of any other nation. Equality points to the particular aspect of sovereignty which implies that if all nations have supreme authority within their territories, none can be subordinated to any other in the exercise of that authority. Nations are subordinated to international law, not to each other. The rule of unanimity is derived from the principle of equality and is responsible for the decentralisation of the legislative function. This rule signifies that all nations are equal, regardless of their size, population and power with reference to the legislative function. The rule of unanimity gives a nation that participates in an international law-giving conference the right not to be bound by the law and the votes of all independent states large and small count the same.

Morgenthau (1968:303–305) adds that sovereignty is not a freedom from legal constraint, not freedom from regulation by international law of all those matters within the domestic jurisdiction of the individual nations, not equality of rights and obligations under international law and finally not actual independence in political, military, economic or technological matters. Inequality of nations and their dependence upon each other has no relevance for the legal status called 'sovereignty'.

Recent developments in international law indicate that the legal status of sovereignty may be changing and may probably not be as clear as Morgenthau seems to believe. It has been argued that state sovereignty is under attack from both procedural and substantive developments in international law. From the procedural side is the rise and recognition of national liberation movements, non-governmental organisations and individual rights through human rights conventions. From the substantive side is the recognition of the right of self-determination outside the traditional colonial context and the recognition of a right to secession. International law did not formally recognise secession as a viable exercise of the right to self-determination until the European Community recognised Croatia and Slovenia on January 15, 1991 (James M. Cooper, 1993). The right to secession is still unclear despite the events in the former Yugoslavia. An unilateral right of a region to secede is not accepted by the large states as was made clear long ago in the American Civil War. The United States considers the present secession war in Chechenya to be 'within well-recognized legal rights in trying to prevent the secession of a region that is historically part of Russia' (Dorinda Elliott and Betsy McKay, 1995:8).

There are many forces operating in the international system of nation-states that undermine the stability of the present system.[4] But there are several roads open for a peaceful legal development of the international system. There might be a gradual transfer of state power to new supra-national institutions as well as increased autonomy of local governments and regions without a significant breakdown of large nation-states and a resulting increase in the number of sovereign states. Future developments depend to some extent on the strength of secession movements and whether the right to secede will be generally accepted in the international community and how this right will be implemented in practice.

The value of sovereignty for small states

A sovereign state has a certain status in the international community, a status that is valuable in itself. A small sovereign state can participate in international organisations, regimes and the United Nations on a formally equal footing with much larger states. Although the influence of small states may not be significant in many cases each state has one vote in the UN and in many other international organisations. Small states can therefore communicate directly to other states at the highest level and in that way directly promote their interests at the highest levels of the international community. The status of sovereignty therefore confers prestige on small states in the international community and opportunities to participate in decision making at the highest level. Representatives from Iceland and Luxembourg have for example held top positions in EFTA and the EU respectively.

The emergence of supra-national organisations and increased interconnection in the world community will increase the load on the resources of small states if they are to participate effectively in all events in the international community. The cost of representation is a problem for the smallest states. Despite advances in communications it seems that this cost is increasing. In Iceland, for example, the cost of the foreign ministry has been rapidly increasing. One reason is the extra work required to fulfil the obligations following participation in the EEA. This includes, for example, translation of a huge volume of EU documents.[5] The possible benefits in the form of new business and trade opportunities must be weighed against the cost of representation.

It has been argued that regions will be the main players in the global economic competition as many states find it increasingly difficult to represent in a unified policy the economic diversity and conflicts within their borders. Small nation-states which *de facto* are hardly more than regions, will enjoy increased political weight and a competitive advantage in the economic competition with other regions. The status of sovereignty gives a small state a competitive advantage over other regions competing at the decision-making level for economic and political power in the European Union (Ágúst Þór Ingþórsson, 1993). Sovereignty creates a privileged position for small states in comparison with peripheral regions, special status areas or even states in a federal system.

Changes in the status of independence

There are probably 'diminishing returns' to sovereignty. If the number of independent states increased greatly (the present number is around 170) and the size distribution of states became more equal the influence of each state would necessarily diminish. The large states in the United Nations could demand increased voting power to avoid being overrun by, for example, 200 small states. A greatly increased number of small independent or semi-independent states makes decision making on an international level more difficult and complex compared to the highly bipolar system of the Cold War. The global system might therefore become more unstable.

The history of the Soviet Union showed that this large state was able to promote some development in many regions characterised by a long history of ethnic and religious conflict.[6] Fragmentation and the disappearance of a strong central government in these areas led to the formation of independent states some of which have since been in a state of constant internal and external conflict. It is not clear whether the process of fragmentation and state formation in the former socialist states is near to completion but it seems to be difficult to stop this process once it has begun. Fragmentation of an economically and territorially integrated region into many small independent states that have no EEZ or other natural boundaries may not be the best solution to political problems. Trade, services and transportation are reduced and economic development is hindered. Cooperation is sometimes impossible because of ethnic and religious conflicts. In that case independence and separation of conflicting parties is the only solution.

In the sixties and seventies the international community did encourage small former colonial states to become independent even if their population was small.[7] The question is whether a further increase in the number of independent states and semi-independent states is to be expected due to secessionist tendencies or otherwise. Two basic factors explain the widespread tendency towards greater self-determination both within and without the former Soviet Union: 'First, there is widespread dissatisfaction with the inefficiency and remoteness of the large-scale, bureaucratic state. Second, there seems to be a growing willingness to question the assumption that inclusion within large states is necessary for security' (Allen Buchanan, 1992:362). In response, liberal democratic theory must be augmented so as to articulate the scope and limits of a right to secede. International law, which only recognises fully sovereign states and individuals within them, must acknowledge new forms of political association, exhibiting a range of types and degrees of self-determination (Buchanan, 1992:351).

An increased number of special status regions covering the whole spectrum from local governmental status to some sort of full independence will complicate the international political system in the same way as the increased number of fully independent states. One problem, for example, is how the new special status regions can participate in international law making. Another problem with Buchanan's idea is that there is no reason why the step towards full sovereignty would not always be taken by regions closest to the status of full independence. The only reason for not doing so is if an autonomous region can enjoy the benefits of association with a state without the costs of sovereignty. This view had strong support in the Faeroe Islands and led to irresponsible financial management which caused among other things the financial collapse of the islands in the early nineties according to Jónsson (1994).

The liberal point of view towards independence that prevailed during the sixties and seventies simply does not apply when it comes to the question of secession of regions in the old states. Russia is resisting a secession movement in Chechenya. In other parts of the world attempts to secede have also led to civil wars as in Zaire (Congo) and Nigeria in the sixties and in Ethiopia where Eritrea seceded successfully and gained independence in 1993. Turkey and Iraq have for a long time resisted independence movements in the Kurdist regions. Regions in India and Africa might possibly try to secede in the nearest future. Secession movements have support in Scotland and probably some other regions in Western Europe but some form of decentralisation or home rule might ease the independence pressures here. Whatever the outcome of these cases, a significant increase in the number of independent states is unlikely while the large states resist most secession movements.

Real independence

Whether any one state is really independent is a difficult question that has many aspects and to which it is probably not possible to give an objective answer. Although a state has obtained international recognition as a sovereign independent state, its freedom of action and its scope of relations with other states may be limited by several factors that are mostly unrelated to the formal international legal status of the state. Therefore it is possible to distinguish between formal independence and real or actual independence. The dimensional approach to the question of independence and sovereignty contradicts the legal conception of independence which is not a matter of degree.

Legal, cultural and economic independence

The legal concept of independence implies that the supreme legal authority is in the hands of the state. In a complex world a small state must often use, with appropriate changes and adaptations, laws and regulations made by larger states especially in the fields of trade and technical matters. Such practical approach to law making does not reduce the real independence of a small state as long as the national assembly retains the power to change the laws. The same goes for international treaties and international laws that are adopted even if entering such agreements means that judiciary powers are transferred to international or external courts. Special agreements between a small state and a multinational firm that intends to invest in the country are not unusual. For example the Icelandic government and Swiss Aluminium agreed upon a special tax treatment for an aluminium smelter which Swiss Aluminium has operated since 1969. It is also debatable to what extent entering a regional association such as the European Union or the European Economic Area does reduce the real independence of a state, large or small. In the case of the EEA for example, Iceland has to adopt many laws and regulations of the EU but as long as Iceland retains the right to withdraw from the treaty, real independence is not sacrificed.

It is difficult to define the term cultural independence although it is doubtless an important part of real independence. Iceland has a rich cultural heritage based on the Icelandic language which has remained almost unchanged (in writing) since the settlement of the country in the ninth century. The cultural heritage was the basis for Icelandic nationalism and an important factor in the fight for independence. The colonial states brought their language and customs to the colonies and it is possible that the reason why some French speaking small island states have chosen to remain at the status of overseas departments or territories is the cultural influence of France in these territories. The pressures may be mounting on the cultural heritage of many small states because of the spread of international mass media through modern communications. A small state might find it challenging to support and cultivate its cultural heritage and stay culturally independent.[8]

Economic independence is an important part of real independence. Five conditions for economic independence are listed below. Fulfilment of these conditions may also be looked upon as desirable goals of economic policy of small states that wish to increase their independence.

1   A single foreign state, institution or a multi-national firm should not dominate any important part of the economic system (in terms of value added for example) through ownership of factories, shops or banks.

2   The most important natural resources of the country should not be in the hands of foreigners, neither private firms nor official institutions.

3   Foreign debt should be within serviceable limits.

4   The export industry should not be based entirely on a single export product.

5   The state should not be dependent on one export market.

A small developing country may lack the basic apparatus to supply its citizens with functions or services normally carried out by the state in other countries. Such services might by supplied by a foreign country which then obtains indirect political influence in the small country. Foreign ownership or control over natural resources is quite common in the developing countries. In a small state with a limited number of natural resources a foreign firm that uses possibly the only natural resource in the state is bound to obtain powerful position in the society. Foreign investment should, however, not be looked upon as an alternative to independence. The main thing for the small state which negotiates with a foreign firm is to make the firm subject to its national laws in principle although some special exceptions may be necessary and the partners agree to settle disputes in international courts.

A state, like an individual or a company, that collects more foreign debt than can be served with its income constrains its freedom and ability to implement economic policies. Foreign debt is a problem for many developing countries. In a small country the state is usually the only acceptable guarantor of foreign loans also for the private sector. Irresponsible financial management might quickly lead to trouble for small states as was the case with Newfoundland in the nineteen forties and more recently the Faeroe Islands.

The fourth and fifth conditions are often referred to as a concentration in production and trade and will be discussed more fully in the next chapter.

Within each state there are often a number of enclaves or special status areas that are not ordinary parts of the political system although they are not independent in international law or international practice. Moreover several small states have obtained home rule but are freely associated to other countries or entrust several functions of the government, especially foreign matters, to other states.

Special status areas and states provide an alternative to independence and integration. They differ greatly in their political and economic status. For analytical purposes they can be located on a continuum, 'which ranges from complete assimilation as a part of a state to a complete independence recognized in the international system' (Smith and Stanyer, 1980:91). The idea of using the independence/dependence continuum obtains further importance when one extends the concept of independence not only to formal status in the international system but also to the ability of a state to implement its own policy in foreign and economic matters without an interference from another state.

The first step in constructing an independence/dependence continuum is to locate states and territories according to constitutional or formal legal status. Recent events in world politics may lead to an increased number of states or regions that do not have a clearly defined formal status. This might be the case with the states within the former Soviet Union and regions in the former Yugoslavia. In most cases there is little doubt about the placement of states and territories on the independence/dependence continuum according to formal status.

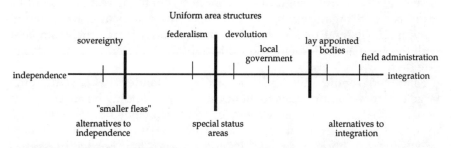

Figure 4.1   The independence/dependence continuum
Sources: Jeffrey Stanyer, direct communication in 1994 and Brian C. Smith and Jeffrey Stanyer (1980:90–94).

Figure 4.1 shows a simplified independence/dependence continuum. The diagram depicts types of area authority characterised in terms of autonomy. The degree of autonomy is a judgement or verdict, resulting from the weighing up of a range of factors. In general it is assumed that the more autonomous the status the more advantageous the situation for the area in question.

Placement of states on the continuum according to the degree of real independence is the most interesting problem. If the focus is, for example, on economic status in international relations, legal status is not the main criterion. Independent states, small and medium sized, compete internationally with regions in large states, federal states or other special status areas and region states. A region state is a part of a natural economic zone that may be composed of regions of more than one nation plus one or more (small) independent states. The primary linkages of region states tend to be with the global economy (see K. Ohmae 1993:78–87).

It is also possible to use the concepts of strength and weakness to locate states on the continuum. This will be done in the next section.

The theory of weak states

An alternative to the idea of real independence is Handel's (1990) theory of weak states. The concept of weakness and strength of a state is fundamental in his study of the behaviour of states in the international system. The terms 'weak state' and 'strong state' are used instead of the usual classification into large and small states. In his theory weak states are supposed to include all the ministates or small states. He is mainly concerned with questions of the political and military power of states although he also discusses the economic power of a state. Four elements determine the strength of a state. They are geographical character, material resources, human resources and organisational capabilities. The weak states are economically, militarily and politically at a serious disadvantage when compared with the great and middle powers although they can find ways to compensate for their weakness.

Main geographic advantages according to Handel (1990:76) are: First, large territory that provides higher probability of finding balanced variety of natural resources and more room for strategic manoeuvre. Second, easy-to-defend borders and difficult terrain. Third, isolation (an island) or a few and weak neighbours, preferably non-hostile ('low border pressure'). Fourth, location on the periphery of the relevant system or subsystem or a non-strategic location.

Handel (1990:70–76) argues that a small territory usually implies fewer natural resources. The weakness of small economies bears a direct relationship to their greater dependence on the importation of foreign raw materials. A small geographical size also has important implications for the military and a strategic situation as the states are vulnerable to surprise attacks and have no 'strategic depth'. A large territorial space also creates difficulties for a small population to defend the boundaries of the country. A geographical location is important as states at the periphery of the international system are in a better position than those located at the centre or 'in the way' of the great powers. The strategic significance is important for other states as the inability of a weak state to prevent occupation by a second power may pose an indirect threat to others. Military strength combines to some extent all the four elements of internal strength.

Handel (1990:76–77) supports the view that small states must defend themselves with an army of their own as much as possible so that they do not create a power vacuum:

> In the past, weak states have on occasion given up hope of defending themselves by their own efforts and have decided to invest only a minimum effort in their military forces [...] Such an attitude is a dangerous one for the weak state. In times of tension and war, a neighbouring great power may fear that the weak state, being a power vacuum in military terms, will fall an easy prey to the great power's enemies - and thus serve as a base against it.

To this may be added that armed forces are also needed for internal security reasons in many states.

Handel (1990:83–90) also points out many difficulties that arise when a weak (small) state tries to establish a strong army. A small population requires a relatively large portion of the population to be mobilised. There is always a difference in the proportion between a mobilised strength and a potential strength between a small state and a large state. High absolute R&D cost forces the weak states to import weapons and that makes them dependent on the seller or they fall victims to a pre-emptive supply. Handel (1990:104–107) then criticises the idea of a passive resistance of small states which can be an important addition to the arsenal of weak states under favourable conditions, but never a substitute for internal military strength or external support.

External sources of strength are needed because internal sources are not sufficient to withstand the external pressure of great powers. This can be done through formal or informal alliances with other states. The simplest way to strengthen the commitment of a great power is to sign a formal treaty or to obtain clear and unambiguous promises of support in case of military attack. But this is not easy if the weak state faces imminent danger. The weak state must use additional means: First, clarify the verbal commitments bilaterally or unilaterally. Second, appeal to the public opinion in the strong state, namely, make it a domestic issue. Third, induce the great power to station troops and maintain military bases on its territories.

Fourth, try to establish a symbolic value by reflecting a positive image of the great power (Handel 1990:122–127). There are certain dangers involved in this strategy. First, the great power may not be ready to terminate the alliance when the weak state so wishes but this danger is thought to be less real when the value of permanent bases has diminished. Second, the weak state could become involved in the conflicts of the great power or be on the target list of the enemy. Third, there is a danger to the cultural integrity of the weak state (Handel, 1990:127–129).

The ability of weak states to turn to other states is a function of the nature of the particular international system in which they operate. In a balance of a power system the position of weak states is affected by three factors according to Handel (1990:187): First, if the great powers are almost equal the importance of the weak states increases in times of tension or conflict. Then the support of weak states becomes more valuable. Second, when the system prefers status quo or when no agreement can be reached between the great powers, they neutralise each other and their mutual jealousy strengthens the position of weak states. Third, the weak states tend to pursue a destabilising policy by supporting the stronger side.

Free riding or the importance of being unimportant

Handel (1990:148–152) points out that weak states enjoy certain advantages as free riders in matters of security. In international relations, collective good accrues to weak states from bilateral or multilateral alliances. The small states enjoy the collective good originating from the direct or indirect protection by the great power. He cites the case of Iceland which spends nothing on defence but NATO guarantees its security in return for a few important military bases.

The Olsonian analysis of alliances and economic communities confirms 'the importance of being unimportant'. Small states do not often make an equitable contribution to the creation of global collective goods, for instance organisations within the United Nations (UNESCO;WHO), NATO and GATT. Mancur Olson (1965) bases his argument on the generalisation of an economic argument and the definition of a collective good. From this analysis he finds that in groups of greatly different size or interest in the collective good there is a tendency towards an arbitrary sharing of the burden of providing the collective good. 'Once a smaller member has the amount of the collective good he gets from the largest member, he has more than he would have purchased for himself, and has no incentive to obtain any of the collective good at his own expense. In small groups with common interests there is accordingly *a surprising tendency for the 'exploitation' of the great by the small.'* This helps to explain the apparent tendency for large countries to bear disproportionate shares of the burdens of multinational organisations, like the United Nations and NATO and some of the popularity of neutralism among smaller countries (Olson, 1965:35–36).

This principle of the 'importance of being unimportant' can also be demonstrated in international trade theory where it can be shown that a small nation (whose output is too small to affect prices) that trades with a large nation gains all the benefits from its own trade.[9] Lack of bargaining power in matters of trade might thus be somewhat mitigated by this principle of unimportance.

Small states as weak states

Handel (1990:48) claims that all the criteria that apply to weak states apply methodologically to ministates except that the question of self defence is irrelevant as no ministate can effectively defend itself by its own devices against even the weakest of other states.

Internal strength of small states with a population of a few hundred thousand depends somewhat differently on geographical position than it does generally in weak states. An absolutely small area of a state precludes many natural resources. Only one natural resources such as oil is often the basis of the economy of such a state. An extreme case is the state of Nauru that is made almost entirely of a phosphate rock and has a population of around 9,000 and area of 21 sq. km.[10] Special resources have sometimes been the basis, at least partly, of independence or secession movements (Katanga, Biafra) which were not accepted by the ruling government and therefore their existence prevented independence in a sense. The great powers might consider a resource such as oil to be of a fundamental importance to their own economy. Such a situation might limit the freedom of a political action by the small state that owns the resource. This situation applies probably to states in the Middle East. A large area also makes the regional problem more difficult for ministates as they have difficulty in keeping a balance in the settlement structure and the cost of infrastructure is relatively much higher than in more densely populated regions.

The small island states control extremely large areas of territorial waters after the recognition of a 200 mile Exclusive Economic Zone. This makes their strategic situation more important and might eliminate their advantage of being geographically remote especially in the future when the resources of the sea will probably become more important. Administration and the control of a large sea or land area is costly and puts extra pressure on administrative resources of small states, especially of developing small island states.

History shows that the most important source of internal strength in a state is the unity of the people. Except in times of total war it is difficult for even the great powers to gain influence within a small state unless internal conflicts are severe. The question then arises whether the political system in a small state is less effective in resolving internal conflicts than the political system in a larger state.

In a small state similar sources of social conflict seem to exist as in other countries such as differences in culture, race and wealth. Such differences have, for example, put a strain on the political system in the Fiji Islands while another example of a severe conflict, about questions of independence, is New Caledonia, an overseas territory of France.[11] Ethnic and cultural homogeneity is probably the most important factor for internal strength and this factor is not dependent on the size of the state.[12]

In figure 4.2 the part of the independence/dependence continuum which includes independent states has been reconstructed and extended according to 'strength' and 'weakness' of states. It is assumed that real independence increases with increased 'strength'.

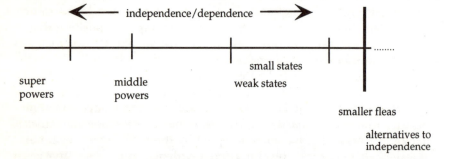

Figure 4.2   The independence/dependence continuum extended

The question is where the small states should be placed. Although the Cold War has ended there are still many unsolved problems in international relations and it is not clear whether general security has increased in other parts of the world than Western Europe. Political, economic and military power is therefore fundamental in determining the position of states on this continuum. Small states will probably always be powerless from a military point of view and dependent on other states for security. There is little doubt that small states are to the right in the part of the 'weak states' on the independence/dependence continuum and a long way from optimum which is at the super powers' end. In a peaceful world where economic viability and the effective size of the state is more important than military strength, many small (island) states would move to the left nearer to the 'weak states'.

It must be admitted that no small state can operate an army that would deter anyone from an attack not even from other weak states at the present level of technology. It is, however, not impossible that a 'doomsday' weapon could be invented which was so simple and cheap to build that even a small state could obtain it. In such a case the world system would change and traditional military strategic thinking would have to change. The small states without any sort of military or para-military units face a danger that other weak states do not have to deal with. A small group of terrorists could invade a small state and do much damage or stage a coup. This happened, for example, in the Comoros in 1978.[13]

In small states the defence of the territory must not be confused with the defence or survival of the population. Iceland, for example, has a defence treaty with the US and a defence force is stationed at the Keflavík naval base. It is probably true that any invasion or attack on the country would eventually be repelled and no part of the country could be occupied for any length of time. The base at Keflavík, however, could be a target in an atomic war and that could result in more than half of the population being wiped out in one stroke as more than 50% of the population live within a distance of around 50 km from the base. Fighting with conventional weapons around or in the country could easily wipe out a large part of the population. In the case of larger states such an extermination of the population in one stroke is unlikely or impossible. The small state must think more of defending the population than defending the land.

An evacuation is sometimes considered and used in times of danger in small states. In 1793–94 Iceland suffered terrible volcanic eruption and earthquake which led to the death of around 20% of the population and an even larger portion of the livestock.[14] It was then seriously suggested to transport the rest of the Icelandic people to Jutland, Denmark. Aldeney in the Channel Islands was evacuated during World War II. A more recent example is the temporary evacuation of Tristan da Cunha in 1961 and also the temporary evacuation of the Westman Isles (off Iceland) in 1973, in both instances after a volcanic eruption.

Iceland's defence policy

Iceland declared permanent neutrality when independence from Denmark was obtained in 1918. This neutrality was not based on military strength as is the neutrality of Sweden for example. Iceland has no military forces of its own.

In May 1940 Iceland was occupied by British troops. This was not surprising at the time as the geographical position of Iceland was vital to the defence of the North Atlantic area. A year later the United States assumed the military protection of Iceland for the duration of the war in accordance with an agreement with the Icelandic government. After the war all troops were to leave the country. With this agreement Iceland ended the declared neutrality policy and accepted subsequently to become a founding member of the North Atlantic Treaty Organisation (NATO). Iceland's contribution to NATO was to grant her allies, in the event of hostilities, similar facilities as were provided in the 1941 agreement with the United States.

After the outbreak of the Korean war and other events the Icelandic government concluded that for security reasons it would be necessary to establish defences in the country. A defence agreement was concluded between Iceland and the United States in 1951 under the auspices and within the framework of NATO. This 1951 Defence Agreement with later annexes is still in force.

This security solution has been a source of a debate in Icelandic politics especially the defence agreement with the United States and the military base at Keflavík airport. In 1956 and 1971 the termination of the agreement was an election issue but failed to generate the necessary support of the Parliament and the electorate. The Icelandic government justified the defence agreement with the United States as a necessary security measure. Discussion of the economic importance of the base at Keflavík to the economy was usually avoided. The economic factor is, however, of great significance. Around 1,600 man-years are registered in direct and indirect services to the defence force and the base provides more than 8% of the export income.[15] In the midfifties, when Iceland experienced a downturn in other activities, this ratio was much higher.[16] The security question turned recently into an economic issue publicly. After the collapse of the Soviet Union the United States proposed to reduce its force at Keflavík. Icelandic authorities have tried to delay such reduction although the security grounds for keeping the defence force have clearly changed.

In terms of security matters it may be argued that Iceland has been a free rider in the North Atlantic Alliance. Instead of being burdened by military expenditure as other NATO allays, Iceland has received considerable economic benefits from the defence agreement with the United States. But the participation of Iceland in NATO has also undoubtedly added to the collective security of the North Atlantic region.[17] A pro-Soviet government in Iceland would have created great difficulties for NATO and probably a sort of 'Cuban' situation.

Iceland has sought a solution to the security problem of small states by entering a defence alliance. When a relatively secure military position is combined with the high level of economic development that has been achieved and the successful participation in various international organisations such as the United Nations, OECD, EFTA and GATT, it can be concluded that Iceland has mostly eliminated the problems related to a lack of political and military power and obtained a high level of real independence. In the next chapter vulnerability related to economic factors, namely an unstable economic growth and foreign trade based on a limited number of export goods and export markets, will be discussed.

Conclusion

The status of sovereignty confers on small states some prestige in the international community and opportunities to participate in decision making at the highest level of the international community. Independence is, however, not real if a state is highly dependent on other states for security or financial support. Use of the independence-dependence continuum clarifies the concept of real independence and shows how the status of small independent states can be compared to special status areas and regions in other states. A small sovereign state has certain political advantages compared to regions within larger states. These advantages can be used to strengthen the competitive position of small states versus peripheral regions or special status areas in international trade.

The theory of weak states is useful to answer questions on small state security and political vulnerability. Small states have no significant armed forces and lack political and economic power. They are therefore suboptimal from the point of view of security. The best strategy for a small states is probably to enter bilateral or multilateral defence alliances. Then the small state enjoys the collective good originating from its direct or indirect protection by a great power. There exists no such easy solution to the problem of economic vulnerability. Dependence or lack of real independence clearly exists in matters related to international trade and economic growth of small states. This is the subject of the next chapter.

Notes

1   Questions about the origin of the state, the justification of state power within countries, state making and state maintenance will not be of concern here.

2   Dommen (1985:4–5) provides examples of 'states' with no territory or no permanent population.

3   Morgenthau seems to use the term nation as the term state is used here. Nation and state are however different sorts of concept. Nation means in the simplest terms individuals with characteristics in

common and State an organisation separate from the individuals within it or subject to it. In ideal-type analysis a nation-state is one in which 100% of the nation live within one state and 100% of the state's population belong to one nation. Such pure nation-states are hard to find although Iceland and Portugal are close.

4   Kennedy (1994:122–134) discusses the future of the nation-state and argues that there are pressures for a 'relocation of authority' because the Nation State is the wrong sort of unit to handle many important problems especially problems on a global scale.

5   Expenditure related to external affairs increased by 67% from 1985–1993 (Þjóðhagsstofnun, 1994, direct communication).

6   Although books from the Soviet Union probably exaggerate the progress made in the Asian republics as, for example Nikolaj N. Mikhaílov (1962:130–157), substantial development occurred in transportation, growth of cities and industry after World War II.

7   See for example the UNITAR study by Rapaport et al. (1971).

8   For a discussion on cultural viability see Doumenge (1985:87–96).

9   See for example Dominic Salvatore (1990:37–38).

10   The Nauru government has invested royalties from the phosphate industry in real estate in Australia and hotels in several Pacific islands. The Nauruans have thus secured a substantial income for themselves after the phosphate has been exhausted (Dommen and Hein, 1985:175).

11   After conflicts between indigenous Kanaks (Melanesians) and white immigrants, it was agreed (the Matignon agreement) in 1988 to divide the country into three regions with Kanak majority in two, white in one. France would rule directly for one year and a referendum on independence will be held in 1998 (Calvocoressi, 1991:165).

12   Japan is an example of a large ethnically and culturally homogeneous state. Iceland is an example of a small such state.

13   The Comoros Islands became a base for right-wing mercenaries after a coup in 1978 with support from France and South Africa, a similar coup by white mercenaries, backed at first by South Africa, in the Seychelles failed in 1980 and 1983 (Calvocoressi, 1991:630–631).

14   Population was estimated to be 48,900 in 1783 but 40,623 in 1785 (The Central Bank of Iceland, 1987:28).

15   Net income from the defence force was Ikr 9,757 million in 1992, total exports Ikr 121,248 million, source: Seðlabanki Íslands (1994). Man-years in direct service were 1,024 in 1991 (Þjóðhagsstofnun, 1994). To this number must be added the employment provided by the building contractors and various services. In 1980, around 580 additional man-years were registered (Framkvæmdastofnun ríkisins, 1982:22). This number has probably not changed much since then.

16   In 1953–1955 at least one fifth of export income originated at the NATO base, see Hagfræðideild Landsbanka Íslands (1960:87–88).

17   The only serious international disputes of Iceland since the war have, however, been with fellow members of NATO, namely the United Kingdom and Germany, in the 'cod wars' and since 1994 with Norway over fishing rights in the 'Loophole' and the Svalbard Islands.

5 Economic characteristics of small states

Small states are suboptimal as autarkic economic systems. A small and often scattered domestic market limits the division of labour and the development of "roundabout methods" of production.[1] Foreign trade provides an escape from the limitations of a small home market, but at the same time flexibility and providence are required ("lead" and "gain" to use Deutsch's terms) to solve the problems that competition in foreign markets raises. The role of the domestic sector is considered to be of a minor importance compared to the foreign sector in small states. It is stated that the domestic sector is confined to a quartermaster function, providing *inter alia* wage goods and services to local workers or inputs into export industries, government services, transport, construction and other support services. These activities exist only because of the market that is created by the export activities or by export income of workers (Dommen and Hein, 1985:152–153).

The following characteristics are often associated with the foreign trade of small states.[2]

1   A high ratio of foreign trade to the national product.

2   A high degree of concentration on a few export products.

3   A high degree of concentration on a few countries to which exports are sold.

A fourth characteristic is sometimes added, namely a low degree of concentration by commodity (and country) in imports. This characteristic will not be of much concern here, the first three have clearly more important implications for the economic status and system capacity of small states.

Studies of economic characteristics of small states

Several studies have been conducted to test the above conjectures. The results have been inconclusive in the sense that 'smaller' states have not been found to be very different from larger ones in terms of these characteristics.

First, Lloyd (1968:24–26) used multiple regression to 'explain' trade ratios. The results were negative in the sense that only 27% of the variation in trade ratios for sixty countries were 'explained' by those variables.[3] Commodity concentration using Hirschman's index in export trade was also investigated by Lloyd (1969:27–28).[4] Multiple regression on the same variables as used in the study of trade ratios did not account for more than 44% of the total variation in commodity variation for his sample of 60 nations. Lloyd used criteria for smallness that allowed countries with population up to 20 million to be included in his sample of small states. His sample was based on information for 1963–1964. Only 5 countries with a population of less than one million were included. Lloyd (1968:35) concluded that '[...] size of a country may play a part but not a dominant part in the determination of trade ratios, commodity and geographic concentration of export trade. Many small countries do not in fact have the characteristics that have often been ascribed to them because the significance which country size may have is swamped by the influence of location, market access, historical ties, the level of the exchange rates, and many other factors'.

Second, Nadim G. Khalaf (1971) made an extensive statistical study on the relation between population, GNP and several trade and income ratios for sixty states to test the impact of size on economic stability, diversity of economic structure, dependence on trade, rate of economic growth and the level of economic development. He failed in most cases to find a strong correlation between his size variable, namely population, and his other variables. In some cases, however, his results were fairly significant, for example a higher ratio of trade to GNP as the size of the population decreased. His sample of states included several 'micro' states although few states with population less than one million had gained independence at the time.

A high ratio of trade to GNP in small states is usually explained as a result of concentration in production and economies of scale in production. Few natural resources mean few or only one export industry that exports the bulk of its products because it cannot find an outlet at home. Competition in foreign markets will lead to an optimal size of plants. Therefore it will lead to even further one-sidedness of the economy and higher trade ratios for a small economy when a small country specialises in the production of goods that are produced under increasing returns to scale.

S. Kuznets (1960:16–20) observed a rise in the ratio of foreign trade to national product as the population size decreases. He also explains this tendency as a consequence of a greater concentration of the economic structure of small nations where a small nation is defined as a nation with a population less than 10 million. He gave three main reasons for expecting a concentration in the economic structure of small states: 1) The result of small size in terms of the geographical area and its limiting effect on the supply of natural, non-renewable resources. 2) The conflict between the minimum or optimum scale of plants for some industries and the limited domestic market of small nations. 3) Small nations may have a marked comparative advantage in a few resources and concentrate upon them to the point where little labour force and a limited number of other resources are left for other domestic production. The tendency of small nations to concentrate production on a limited range of activities implies that foreign trade is of greater weight in the economic activity of small nations than in larger nations. This applies especially to small developed nations that have attained fairly high levels of per capita output and consumption. There the variety of goods demanded by ultimate consumers is far wider than domestic output of final goods. Small developing countries also tend to have some comparative advantages and the weight of their exports in relation to total activity is likely to be greater than in larger equally developing countries.

For most products, for example food and other consumption goods, the extent of the market is directly proportional to the number of consumers and their income and spending habits. A small population size is therefore equivalent to a small market. Foreign trade would largely eliminate the effects of a small domestic market but transport or transaction costs will in some cases also limit the size of the export market. Such costs are dependent on geographical factors like location or distance from other countries. During this century transport costs have been diminishing greatly and are now well below 10% for most manufactures. In the case of bulky goods such as coal or perishable products such as milk transport costs are higher than for manufactures.[5]

Many small states are abundant in natural resources within their boundaries or EEZ. Natural resources form the basis of a comparative advantage in international trade for many small states but there are other important economic activities not related to natural resources that form a basis of foreign trade. This is the case with service industries such as finance, insurance and tourism. Minerals and raw materials can also be imported and processed for re-export. A narrow resource base is therefore not the only explanation of the observed trade characteristics of small states. The reason for a limited range of export goods is more likely to result from the simple fact that a small state with a population of less than

68

one million has at most a work force of a few hundred thousand. Given a similar share of services in GDP as is normal in large industrial states or 50%-60% of the total manpower there is not the manpower available for many export production units given a minimum size of at least 50–100 employees per firm. When a nation has a rich natural resource such as oil or fishing grounds its utilisation requires a great deal of available manpower even though capital intensive methods are used. This means that the connection between efficient size of economic units and population is probably more important in explaining relative one-sidedness of small economies than natural resource endowment.

International trade has expanded since the studies above were done and the number of independent small states has multiplied. Here it will be shown that states with a population of less than one million have generally the abovementioned theoretically expected characteristics. Iceland, in particular, is a characteristic small state in economic terms.

The implications of the trade characteristics for small states are *inter alia*:

1   Fluctuations in export earnings.

2   Unstable terms of trade.

3   Unstable exchange rates and high inflation.

Fluctuations in the price of primary commodities is a well known phenomenon. It is therefore to be expected that states that export primary products, mainly developing states, experience fluctuations in export earnings. When a small state exports a limited number of primary commodities it is less likely that effects of falling prices in one commodity can be eliminated by rising prices in another commodity. Several studies have confirmed that export earnings of developing countries are unstable and also that instability increases with decreasing size of the state.[6] Fluctuations in export earnings will be taken as given in this study except for export instability in Iceland that will be examined in more detail later. Exchange rates, inflation and the terms of trade will also be discussed later. If negative aspects of those characteristics were prevalent the result would be slower economic growth in small states than in medium sized or large states. This turns out not to be the case. High level of specialisation in international trade does not seem to be a disadvantage for small states. As an example Iceland has obtained a high standard of living despite fluctuations in economic growth and export earnings. International trade clearly increases the effective size of small states and helps to overcome the suboptimal characteristics of a limited home market.

Trade and economic growth of small states

An analysis was undertaken to find out whether small states with a population of less than one million can be characterised by a high ratio of trade to GNP and a high concentration in exports and export markets. Growth performance was also examined. The share of the foreign sector in the national economy is best observed in terms of value added. Information on value added for each industry is not easily available for small states and information is incomplete on trade in services, which is an important source of foreign currency for many small states, e.g. tourism. The available data will, however, provide sufficient accuracy for the purpose of this study.

The share of foreign trade in the national product

In figure 5.1 the average ratio of exports and imports in goods and services to GDP is shown. The states have been grouped into six classes according to population size. It can be seen that the ratio of trade to GDP lowers as population size decreases in these five groups.

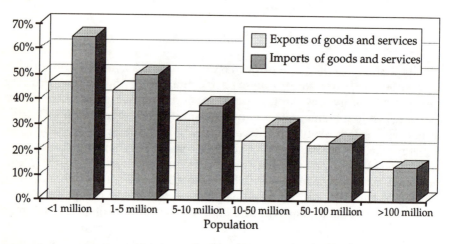

Figure 5.1   **Average ratio of trade and country size, 1988. Exports and imports of goods and services as a percentage of GDP**
Source: UNCTAD (1991).

A similar result is obtained by summing up total merchandise exports and imports for all states in these groups as shown in table 5.1.

Table 5.1

Ratio of merchandise exports and imports to GDP, 1988

Population class	Number of states in each class	Ratio of total exports in each class to total GDP	Ratio of total imports in each class to total GDP
<1 million	32	35%	47%
1 - 4.9 million	37	40%	40%
5 - 9.9 million	18	38%	38%
10 - 49.9 million	40	23%	23%
50 - 99.9 million	10	19%	19%
>100 million	9	13%	15%

Source: UNCTAD (1991).

The following scatter diagram shows the distribution of trade ratios. Calculation of standard deviation confirms that there is a wide variation about the group averages but it is clear that small states generally have a higher ratio of trade to GNP or GDP than larger states. The ratio of imports to GDP is shown instead of exports because in the case of a number of small states exports of goods and services are very limited but aid and transfer payments seem to fill the gap between exports and imports. Imports therefore show better the importance of the foreign sector.

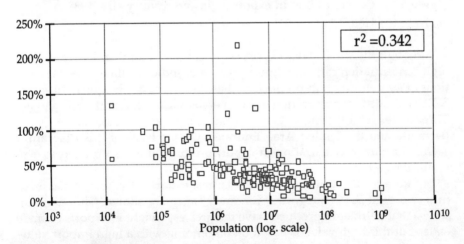

Figure 5.2   **Ratio of trade and country size, 1988. Imports as a percentage of GDP**

Source: UNCTAD (1991).

71

The following diagram shows that the average commodity concentration index decreases with population size and the number of goods exported increases, except for the largest states.[7] The concentration index or ratio applies only to merchandise trade so that important service industries such as tourism are excluded. The concentration index is widely scattered around the mean so there is no simple connection between a concentration in the export structure and the population size.

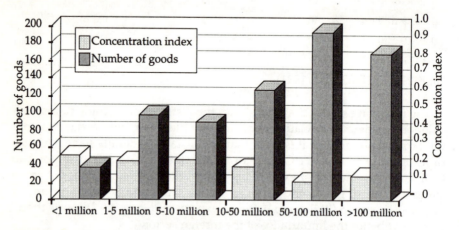

Figure 5.3   Concentration in export trade and country size, 1988
Source: UNCTAD (1991).

Although the small states export fewer commodities than larger states a high concentration ratio in many larger states indicates that other explanations of concentration in production than small size can be given. In figure 5.4 commodity concentration ratios versus GDP per capita are shown. Three states with the highest concentration and highest income are Qatar, Bermuda and the United Arab Emirates. On the basis of this diagram it seems that low concentration in export production is also related to economic development.

In figure 5.5 the values of the two largest export activities (as classified into two digit SITC categories) for small states are shown. This diagram shows that petroleum products have the greatest weight in exports of small states but this high value is based on oil refining with a high import value in unrefined oil. Except for oil, food products, including fish, fruits and sugar, are the main export commodities. This figure captures around 80% of all exports of the states in the sample.

72

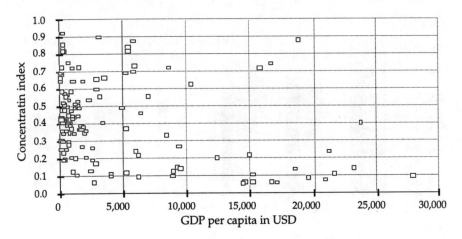

Figure 5.4   Concentration in trade and GDP per capita, 1988
Source: UNCTAD (1991).

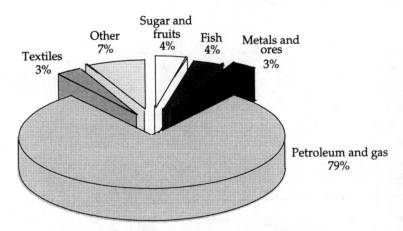

Figure 5.5   Main export commodities of small states, 1988
Source: UNCTAD (1991).

Concentration in export and import markets

In figure 5.6 the geographic concentration of exports and imports is shown. The states have been grouped into six classes according to population size. It shows the share of the two largest partners in merchandise exports and imports. The small states have more geographic concentration than larger states. The ratios are again dispersed widely around the mean as is shown in figure 5.7.

73

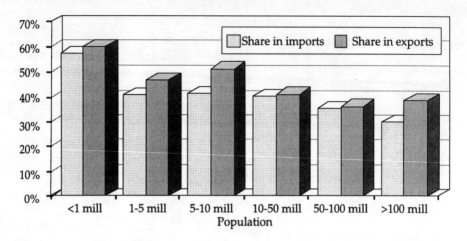

Figure 5.6 **Geographic concentration in trade and country size, 1988. Merchandise imports and exports, average percentage share of two largest partners**

Source: United Nations (1991).

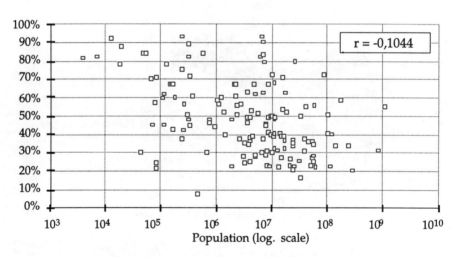

Figure 5.7 **Geographic concentration in exports and country size, 1988. Percentage share of the two largest partners in exports**

Source: United Nations (1991).

The result of this simple statistical analysis is that there is little doubt that small states have the characteristics described at the beginning of the chapter. It is unlikely that results would be much different if more complex methods or more exact data were used. The use of broad groupings of states has, however, been criticised. Lloyd (1968:29) criticises the method of using broad grouping of countries because these groupings have hidden the wide variation about the group averages. He recommends the use of multiple regression that simultaneously tests several variables.

Although it is likely that several variables or factors are needed to explain the variation in some trade characteristics of small states it does not seem necessary to use more complex statistical methods to demonstrate that these characteristics apply generally to small states.

Economic growth in small states

Data on economic growth rates over a long period is only available for a limited number of small states. Comparison of gross national product and GNP per capita between countries can be misleading because of several well known reasons unless measured in Purchasing Power Parity (PPP) where account is taken *inter alia* of the prevailing price levels in the countries. Such calculations are not available in most cases. The available data shows that economic growth in small states has on the average not been lagging behind larger states.[8] Boris Blazic-Metzner and Helen Hughes (1982:87–89) found that small developing states with population below one million (a sample of 21) grew faster 1965–1978 than states in the category one million to five million and often faster than developing states with population over five million. They also found that growth of the population was slower and that growth of GDP per capita was higher in small developing states than larger developing states in the periods 1965–1973 and 1973–1978. They interpreted the result as consistent with the view that a demographic transition from low life expectancy/high fertility to higher life expectancy/lower fertility is faster in rapidly growing countries with equitable social policies, but some small states seem to adopt welfare policies relatively early.

Some of the small states had a serious recession in the 1980s but economic growth has despite this been satisfactory on the average in small states as the following table shows. This table shows that the small states are at very different stages in economic development. Some are among the poorest countries in the world, others among the most prosperous.

Table 5.2
Economic growth in small states

Country	Popula-tion 1988 in '000	Average growth rate of GDP 1960–1989	Average growth rate of GDP 1980–1989	GDP per capita 1988 in USD
Bahrain	481	...	2.0%	6,980
Barbados	254	2.5%	1.7%	6,057
Cape Verde	352	4.5%	0.4%	748
Comoros	515	3.2%	3.4%	402
Cyprus	687	4.8%	5.9%	6,170
Djibouti	387	5.6%	1.6%	589
Equatorial Guinea	336	...	3.2%	438
Fiji	738	4.1%	0.2%	1,457
Gambia	815	4.9%	...	271
Guinea-Bissau	928	2.8%	3.8%	159
Guyana	794	1.5%	-2.1%	454
Iceland	248	4.8%	2.8%	23,851
Luxembourg	371	2.9%	3.7%	18,203
Maldives	199	...	11.3%	...
Malta	349	7.6%	3.1%	5,246
Samoa	167	...	0.6%	680
Sao Tome and Principe	115	1.0%	-3.9%	548
Seychelles	84	3.7%	2.3%	3,511
Suriname	406	2.2%	2.3%	2,886
Swaziland	738	4.9%	3.8%	825
Average of small countries		3.8%	2.4%	4,182
Average of developed countries		3.5%	3.2%	17,387
Average of developing countries		4.9%	2.7%	1,031

Source: UNCTAD (1991).

Development of the Icelandic economy

At the beginning of this century the Icelandic economy consisted mainly of small scale farmers and fishermen. Modern fishing industry was starting to develop thanks to the introduction of the steam trawler and motorboat.

The technical changes in the fishing fleet were a part of a profound change in the structure of the economy that took place during the first four decades of this century. The outstanding feature of that development is an occupational change from primary production to secondary and tertiary activities. In 1890, more than 80% of the population was engaged in farming and fishing, 3% in manufacturing and construction and 6% of the population were engaged in trade, transportation, government and personal services. In 1940, 46% was engaged in farming and fishing, 22% in manufacturing and construction and 27% in trade, transportation, government and personal services (Björnsson, 1967:220). Since then occupational distribution has been characterised by rapid growth of secondary and tertiary activities as in other developed states. Occupational distribution in the year 1991 is shown in the following table:

Table 5.3
Occupational distribution in Iceland, 1991

Agriculture	5.4%
Fishing	5.5%
Fish processing	6.0%
Industry and construction	22.3%
Commerce, restaurants and hotels	14.6%
Communication	6.9%
Financial institutions, insurance	8.4%
Government services	18.5%
Other services	12.4%
Total	100.0%

Source: Seðlabanki Íslands (1994).

Before the 1930s the chief fish export article was salted cod. The most important markets were Spain, Portugal and Italy. Fish on ice was also exported to British and German ports. During the first World War the demand for food increased sharply but soon after the war prices of imports rose faster than export prices. Things were improving when the great depression caused markets in Southern Europe to deteriorate and the British and German markets became subject to quota restrictions. In addition, following the Ottawa Conference of 1932, tariffs on fish in Britain were raised by 10% in the year 1933 (the imperial tariff) which among other indirect effects caused further difficulties to the Icelandic economy. Another set back came when the civil war in Spain led to a total collapse of the most important saltfish market for Iceland. After 1930 new market conditions were met by changes in the fishing industry. Herring fisheries were expanded and factories built for production of herring oil and herring meal.

The first quick-freezing plant was established in the year 1930 and 31 plants were operating in 1940. At the same time agricultural productivity increased but agricultural exports declined from 20% of total exports in 1900 to 10% in 1940. The marketing difficulties of the 1930s disappeared soon after the outbreak of World War II. During the war most of the Icelandic fish products were sold to Great Britain at favourable prices. Foreign debt which amounted to 113 million krónur at the outbreak of the war changed to a net claim of 580 million krónur at the end of the war (Björnsson, 1967:222–223).

After World War II fishing continued to be the dominant export activity. New markets were found in the United States and Eastern Europe and more processing took place both in Iceland and overseas in final processing factories owned by the Icelandic producers. There had been more inflation in Iceland during the war than in those countries that Iceland traded with but the exchange rate had been practically unchanged since 1939. Export earnings dropped due to failure of the herring catch in 1945 and poor herring catches for several years thereafter. Devaluation of the króna was not considered politically feasible. Therefore severe balance of payments difficulties were met by import restrictions and export subsidies. In March 1950 the Icelandic króna was finally devaluated. The budget was balanced and credit restricted and import controls relaxed but due to declining terms of trade subsidies to export industries were soon reintroduced and lasted until 1959 (Björnsson, 1987:223–229).

There was a market increase in economic growth in the first half of the 1960s compared to the preceding decade. This development was due to increased catches especially herring catches and a change in the economic policy. A new stabilisation program of 1960 included the following changes: The króna was devaluated by around 50%, a complicated system of export subsidies was mostly abolished, the rate of interest was increased, control of investment that had been exercised since 1947 was abolished and imports were liberalised (Björnsson, 1967:229–230). In 1967 and 1968 fish catches dropped sharply mainly because the herring disappeared almost completely. This alongside falling export prices led to a slump in the economy. GDP fell by 5.3% between 1967 and 1968 (Þjóðhagsstofnun, 1993). In 1969 catches recovered and the terms of trade began to improve. From 1970 to 1980 the economy enjoyed a high growth rate or over 6% on the average although the terms of trade continued to fluctuate as well as catches. Another natural resource was exploited, namely hydro-power in the production of aluminium and ferro-silicium. The oil price increases in 1973 and 1979 hit the economy hard and the terms of trade deteriorated. The main reason was that the profitability of the fishing fleet was sharply reduced due to the large oil consumption of the fleet. Another slump in the Icelandic economy followed the collapse of the stock fish market in Nigeria along with poor fish catches in 1983. The situation improved again in 1985 and a record growth occurred in 1987 mostly due to high prices of fish and good catches. Economic growth has been slow or negative in Iceland since 1988. Exports stagnated following a decline in fishing stocks and fish catches and a fall in the prices of aluminium and ferro-silicium. The economy started to recover in 1993 by an 0.9% increase in GDP and in 1994 economic growth is estimated to be around 2% (Seðlabanki Íslands, 1994:27).

Trends in economic growth

An analysis was undertaken to investigate quantitative aspects of economic growth in Iceland from 1945. In figure 5.8 a volume index of GDP from 1945 is shown and changes on the previous year. Growth is irregular and slows down after each period of growth lasting between 4–6 years. These cycles are probably not ordinary business cycles but rather a result of poor fish catch due to reduction in fishing stocks or fall in market prices for fish. The slump that began in 1988 is, besides reduced catches and fall in prices, due to internal problems that are related to high inflation and financial mismanagement in the preceding years. Despite the fluctuations the average economic growth from 1945 to 1992 was 4.1% per annum.

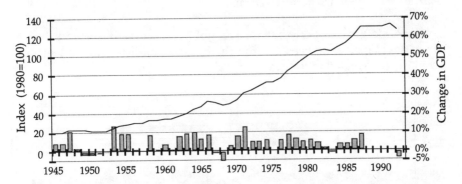

Figure 5.8   Economic growth in Iceland, 1945–1992. Volume index of GDP in Iceland and per cent changes on the previous year

Source: Þjóðhagsstofnun (1993).

Growth of the population has generally been relatively rapid. Population was 143,973 in 1950 but 262,202 in 1992, average rate of growth 1.44%. In figure 5.9 average growth of GDP per capita is shown.

In the first part of the period, up to 1960, growth per head was similar as in other western countries in the lower category. From 1960 to 1989 growth in GDP per head in Iceland was higher on the average than in other developed market countries or 3.6% per annum versus 2.6% (UNCTAD,

1991:434–444). The slow growth since 1988 has caused Iceland to fall behind other OECD countries. In 1991 GDP per head in Iceland measured in Purchasing Power Parity was 17,237 US dollars, $22,204 in USA. $19,500 in Germany, $19.107 in Japan, $17,772 in OECD countries (Þjóðhags-stofnun, 1993:184–185).

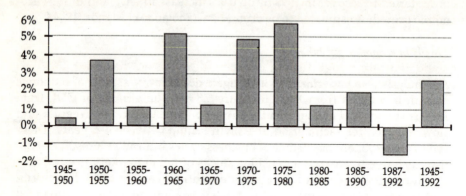

Figure 5.9   **Economic growth per capita in Iceland, 1945–1992. Percent-age changes in GDP per capita**

Source: Þjóðhagsstofnun (1993).

Economic growth performance has not been worse in Iceland than in other developed countries but the fluctuating character of the Icelandic economy is revealed in figures 5.8, 5.9 above and 5.10.

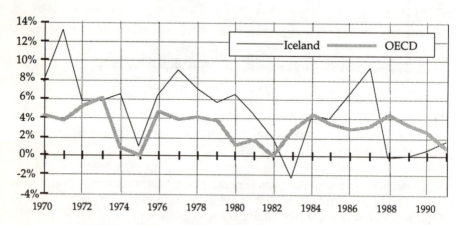

Figure 5.10 **Economic growth in Iceland and OECD, 1970–1991. Annual percentage growth rate of GDP**

Source: Þjóðhagsstofnun (1993).

Have the fluctuations been more severe than in other OECD countries? This question has been answered roughly by simple statistical analysis. Mean annual growth of GDP in Iceland was 4.8% from 1965 to 1988 with standard deviation of 4.2 and coefficient of variation 0.88. For OECD (without Iceland) average growth was 3.5% with a standard deviation of 2.5 and a coefficient of variation equal to 0.71 (OECD, 1990:70). This comparison shows that the economy has experienced somewhat greater fluctuations than larger developed economies in OECD. A more detailed analysis shows that the span of fluctuations is relatively short which should indicate a high degree of adaptability in the economy and the economy has also become more stable in recent years.[9]

The Icelandic economy has been more unstable than other developed economies in the field of monetary matters. The average annual inflation rate was 30.5% in Iceland 1965–1988 but 9.1% in the other OECD countries (OECD, 1990:70). The inflation problem and monetary policy will be discussed in the next chapter.

Growth of exports

Economic growth has been closely correlated with the growth in foreign trade. In figure 5.11 a volume index of GDP, exports and imports is shown. The jumps in imports 1974 and 1987 are associated with a high rate of domestic monetary expansion.

These volume indices show that the export industries have been a leading sector in the economy or at least not a lagging sector despite the fluctuations experienced in export income. The ratio of trade to GDP is lower than the average for small states (see figure 5.1) but there is a reason to believe that the importance of trade to Iceland is underestimated by this measurement. If GDP per capita is converted to dollars in terms of current purchasing power parity, it is lowered by approximately one third from current exchange rates.[10] Export income is largely paid in dollars and it is therefore the domestic sector that is mostly reduced in value by this PPP measurement. This implies that activities which earn foreign exchange are undervalued compared to other activities.

The growth and fluctuations of export income for the period 1900 to 1940 have been estimated quantitatively and analysed by Sigfús Jónsson (1980). He found an average growth rate of 4.7% per annum in 'real export income' where index of export prices was lowered or incremented according to an index of the terms of trade. But real export income varied greatly (average annual variation was 22.5% from the trend line). Variations in export income were related to fluctuations in demand in foreign markets, changes in the terms of trade and variation in the levels of catches and agricultural exports. Correlation showed that 76% of the variation of 'real export income' was accounted for by the variation of the level of catches, measured in cod-equivalent units (S. Jónsson, 1980:191). During the period 1901 to 1940 competition in export markets was characterised by price competition but supply conditions in Iceland as well as output from the competitors affected prices.

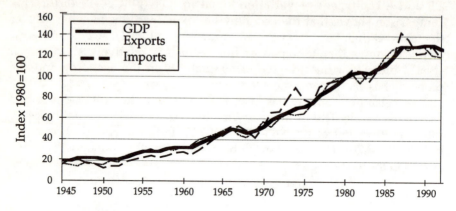

Figure 5.11 **Growth of trade and GDP in Iceland, 1945–1992. Volume index of exports, imports and GDP**

Source: Þjóðhagsstofnun (1993).

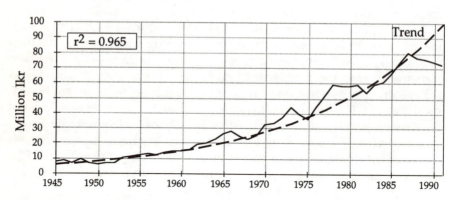

Figure 5.12 **Real export income in Iceland, 1945–1991. Constant 1980 prices**

Source: Calculated from Þjóðhagsstofnun (1991) and (1993).

In figure 5.12 a calculation of the index of real export income from 1945–1991 is shown. The diagram shows a slow and unstable growth of export income first after the war but export income started to grow faster after 1960. In 1967–1968, 1974–1975 and 1982–1983 and from 1988 slumps are clearly visible.[11] The average rate of growth is 6.0% per annum compared to 4.7% in the period 1901–1940.

Service industries have been growing in importance as an export activity. In figure 5.13 exports of services and merchandise exports are shown. Formerly, export services consisted mainly of services to the defence force in Iceland and transport but other services such as tourism, have been growing rapidly.

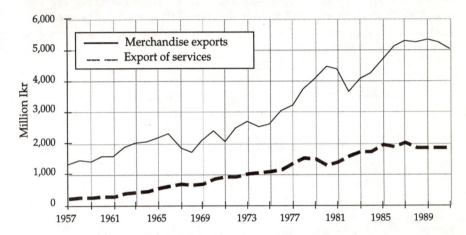

Figure 5.13   Growth of service and merchandise exports in Iceland, 1957–1991. Constant 1980 prices

Source: Þjóðhagsstofnun, direct communication.

Concentration in exports

Exports have been characterised by a high concentration in production and dependence on a limited number of foreign markets. For the first two decades of the century agriculture and whaling constituted around 30–40% of exports and fish products around 60–70%. In 1915 whaling was banned and agricultural exports diminished down to around 10% while fish products climbed to 80–90% of exports in the 20's. Concentration in export outlets was marked as shown by the fact that the four largest export markets took 85% of exports in 1906–1907, 80% in 1913, 79% in 1920, 70% in 1930 and 56% in 1938 (Björnsson, 1967:225). Although four countries bought most of the Icelandic exports (and provided most of the imports as well, 89% of the imports came from four countries in 1913, 78% in 1920, 82% in 1930, 75% in 1938 100% in 1944) these were not the same countries all the time. At the beginning of the century most of imports and exports went through Denmark but Denmark's importance declined gradually until its share dropped to below 10% in the 1930s. Then Spain became very important (36% in 1920, 34% in 1930) but after the outbreak of the civil war exports to Spain stopped completely. Britain has kept a large share most of the time (15% to 20% of exports until 1939) and during the war most of the exports went to Britain (90% in 1944). After the war a dispersion of export markets occurred gradually although the geographical concentration is still marked. In table 5.4 the share of four nations is shown. These nations have most of the time provided the largest export markets. Japan and France have increased their share in exports recently. Japan received 6.0% in 1990 and 7.6% in 1992, France had 9.0% in 1990 and 9.9% in 1992. These nations also hold a large share of the import market but imports are not as concentrated as exports.

Table 5.4
Percentage share of the four most important countries in merchandise exports and imports of Iceland, 1946–1990

Exports	'46–'50	'51–'55	'56–'60	'61–'65	'66–'70	'71–'75	'76–'80	'81–'85	'86–'90
UK	29.9	13.1	10.0	20.1	15.5	10.1	16.8	14.3	21.8
USA	8.9	17.4	13.0	15.3	22.9	27.9	25.8	26.1	15.6
West Germany	10.1	5.4	8.1	9.8	8.5	8.2	8.9	8.5	10.8
USSR	8.0	9.2	18.2	9.2	9.1	7.9	4.9	7.1	3.4
Total	56.9	45.1	49.3	54.4	56.0	54.1	56.4	56.0	51.6
Imports	'46–'50	'51–'55	'56–'60	'61–'65	'66–'70	'71–'75	'76–'80	'81–'85	'86–'90
UK	30.3	16.7	9.9	13.3	13.7	11.6	10.3	8.6	8.1
USA	20.8	20.5	14.7	13.4	11.6	9.0	8.1	7.6	9.4
West Germany	1.0	6.7	10.5	11.9	14.0	11.9	10.5	12.3	14.0
USSR	0.8	5.5	16.9	10.9	7.6	8.7	10.0	8.9	4.5
Total	52.9	49.4	52.0	49.5	46.9	41.2	38.9	37.4	36.1

Sources: Hagstofa Íslands [The Statistical Bureau of Iceland] (1994) and Þjóðhagsstofnun (1993).

In 1992 68.6% of total merchandise exports went to the EU and 48.6% of imports were obtained from the EU.

In table 5.5 exports by industrial origin are shown. Concentration in export production increased after 1945 and until the late 1960s fish and fish products were over 90% of exports. Only after 1970 did manufacturing products become important when an aluminium smelter and other factories started to operate.

Table 5.5
Merchandise exports of Iceland by origin, 1901 to 1990,
percentage of total value

Average for years	Products of fishing	Products of agriculture	Products of industry	Other
1901–1905	59.0	20.9	0.2	19.9 (18.2)
1906–1910	64.2	21.7	0.4	13.7 (12.2)
1911–1915	73.2	23.1	0.2	3.5
1916–1920	74.5	21.4	0.0	4.1
1921–1925	84.9	13.3	0.1	1.7
1926–1930	87.8	11.1	0.0	1.1
1931–1935	89.3	9.6	0.0	1.1
1936–1940	85.3	13.0	0.0	1.7
1941–1945	92.6	6.2	0.1	1.1
1946–1950	90.1	6.3	0.3	3.3
1951–1955	92.8	4.6	0.2	2.4
1956–1960	90.6	6.8	0.2	2.4
1961–1965	91.2	5.8	0.9	2.1
1966–1970	84.9	5.6	6.3	3.2
1971–1975	74.6	3.0	19.9	2.4
1976–1980	73.4	2.3	21.8	2.5
1981–1985	71.3	1.5	24.2	3.0
1986–1990	73.9	1.7	21.4	3.0

From 1901 to 1910 products of whaling are shown in parenthesis in the last column. Their share dropped to zero in 1916 but has been between 1% and 2% since 1946 until 1987.

Sources: Hagstofa Íslands (1984) and Þjóðhagsstofnun (1993).

Export markets have been relatively stable. There are, however, to be found instances of serious blows to the economy that have resulted from a heavy reliance on single markets and/or products. This happened as mentioned before in the thirties when the civil war in Spain caused the loss of the saltfish market there. More recent instances are from the cod wars and an unrest in Nigeria. A ban on landings of Icelandic trawlers in British harbours was imposed at the beginning of the fifties as a consequence of a dispute between Iceland and Britain about the fisheries limit around Iceland. This ban created market difficulties for a most important export product (iced fish) and as a result Iceland had to divert its trade to East European countries where transactions were on a barter trade basis. Exports of stock fish to Nigeria had been growing rapidly for some years in the late 1970s and at the beginning of the 1980s. Then in the year 1983 the market suddenly collapsed, probably due to both economic factors (falling oil price) and political factors in Nigeria (Nigeria had 7.1% of exports from Iceland in 1980, 13.1% in 1981, 3.8% in 1982, 4.8% in 1983 and only 0.2% in 1984). The recent conclusion of the Uruguay Round will lead to freer international trade in agricultural products and fish. This makes it easier for Iceland and other small states that export food products to seek new markets if traditional markets become unstable.

Conclusion

Previous studies of small states, when the sample of small states included only a few states with population less than one million, did not confirm all the theoretically expected economic characteristics of small states. Analysis that includes most small states with population less than one million demonstrates that this group of states has the expected characteristics, namely a high level of trade and dependence on a limited number of export commodities and export markets. The negative implications of these characteristics have not been reflected in the growth performance of small states. Many small states have obtained economic growth rates that are above the average values experienced by the larger nations. Small states must be prepared for instability in export prices and income which results from their dependency on a few primary or semiprocessed export products. This makes it very important to manage the monetary and financial sector of the economy prudently to avoid high rates of inflation and other problems that hinder economic growth. Monetary policy is the subject of the next chapter.

Notes

1   See Allyn A. Young (1969) for the importance of the extent of the market in economic progress.

2   See for example Lloyd (1968:14), S. Lall and S. Ghosh (1982:145) and A. D. Knox (1967:37–38).

3   Lloyd (1968) used total trade (exports+imports) divided by GDP as the independent variable. The independent variables were country size (as measured by GDP), population, area, GDP per capita, percentage of GDP spent on gross capital formation and the degree of industrialisation as measured by the percentage of GDP produced in ISIC groups 1, 2, 3 and 5. Dollars were converted by purchasing power parity exchange rates.

4   For a definition see Lloyd (1968:26).
5   The CIF/FOB ratio for the world has gone from 1.090 in 1950 to 1.056 in 1987 (International Monetary Fund, 1988:128–129). This ratio is a very rough indicator of transportation costs.
6   Helleiner (1982:166–167) and Dommen and Hein (1985:171) discuss several studies.
7   The concentration index is the Hirschman index, see UNCTAD (1991). Number of goods exported at the three-digit SITC (Standard International Trade Classification) level includes products which are greater than $100,000 in 1988 or more than 0.3% of the country's total exports.
8   Ronald V. A. Sprout and James H. Weaver (1992:244) in their study of international income distribution 1960–1987 (using PPP) found no connection between the size of a country and its economic growth rate. Small countries were among the fastest growing countries and none of the largest or smallest countries were among the ten worst performers.
9   According to Seðlabanki Íslands [The Central Bank of Iceland], direct communication.
10   GDP per capita in 1991 was $24,960 but measured in PPP $17,237 (Þjóðhagsstofnun, 1993:186).
11   Exports increased 6.4% from the previous year in 1993 and an increase of 4.9% is estimated in 1994 (Seðlabanki Íslands, 1994:27).

6   Monetary problems and policy

A high ratio of trade to GNP and concentration in production and trade make small states sensitive to price conditions in their export markets. Unstable export earnings put pressures on the monetary system in small states, possibly leading to inflation and frequent adjustments in the exchange rates. The question is whether small states are more prone to inflation than larger states and whether or not small states should use their own national currency and pursue active monetary policies. It turns out that small states are not more liable to inflation than larger states if the currency is stable. The appropriate exchange regime for small states is fixed or pegged exchange rates.

Inflation and exchange rates in small states

In figure 6.1 average annual inflation in a number of small states and the average for the world and high income countries (as defined by the World Bank) is shown. It can be seen that many small states have a low level of inflation compared to other states. Only four states are above the world average and Iceland is one of them.

Table 6.1 below has been constructed to analyse changes in inflation and exchange rates in small states. Small states that peg their currencies have a low level of inflation. Others such as Iceland show a high rate of inflation and unstable currency. The small states in the Organisation of the Eastern Caribbean States[1] which share a central bank and a common currency, linked to the US dollar, provide an example of countries with low inflation rates although they are strongly characterised by high trade ratios, a narrow resource base and concentration in trade.[2] The data in this table indicates that fixed exchange rates are a prerequisite for a low rate of inflation.

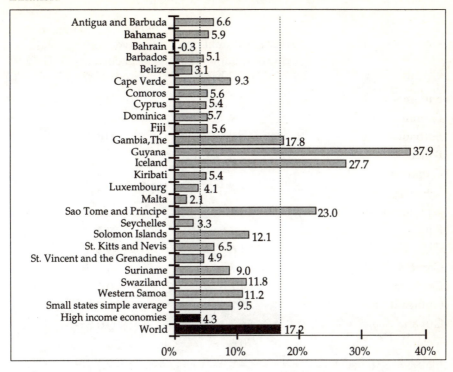

Figure 6.1 **Inflation in selected small states, 1980–1992. Average annual percentage rate of inflation**

Source: World Bank (1994).

The question of floating versus fixed exchange rate in small states is unresolved in theory. R. Mundell (1961:662–663) in his discussion of optimum currency areas produces three arguments against small currency areas.[3] First, increasing cost of currency conversions (with increased number of small currency areas). Second, the speculative argument against flexible exchange rates, the market must be so large that a single speculator can not affect the market price. Third, the assumption that the community is not willing to accept variation in real income through adjustment in money wages, only through variations in the rate of exchange. The necessary degree of 'money illusion' becomes higher in a smaller currency area. This argument suggests that small states are better off by pegging their currencies or by using an 'external' currency. Two additional reasons for preferring a fixed exchange rate policy in small states can be given.

Table 6.1
Inflation and exchange rates in small states, 1987–1990

Country	Average annual percentage inflation rate 1987–1990	Percentage change in the conversion rate of the local currency to USD from 1987 to 1990
Bahamas	4.7	0.0
Bahrain	0.9	0.0
Barbados	4.7	0.0
Belize	2.8	0.0
Comoros	3.3	-9.4
Cyprus	3.9	-4.2
Equatorial Guinea	3.1	-9.4
Fiji	8.7	19.4
Gabon	1.6	-9.4
Gambia, The	10.7	13.3
Grenada	4.7	0.0
Guinea Bissau	58.7	290.7
Guyana	56.2	305.0
Iceland	20.6	50.7
Luxembourg	2.9	-10.5
Malta	1.6	-8.6
Mauritius	11.8	15.4
Sao Tome and Principe	32.2	164.4
Seychelles	2.4	-4.6
Solomon Islands	13.4	26.5
St. Lucia	3.1	0.0
St. Kitts and Nevis	3.2	0.0
St. Vincent	3.1	0.0
Suriname	6.9	6.2
Swaziland	42.1	27.6
Tonga	7.9	-13.9
Trinidad and Tobago	10.1	18.1
Vanuatu	7.1	6.1
Western Samoa	10.0	9.4

Source: The World Bank (1992).

The first reason is that frequent devaluations have strong inflationary effects in small countries with a high ratio of trade to GDP as the case of Iceland confirms. A fixed exchange rate is a necessary condition, but not a sufficient condition, for a low inflation rate. The second reason is that devaluation will probably not lead to an increase in exports or larger market share abroad. This is due to the fact that merchandise exports of small states are generally resource based and therefore supply constrained (decreasing returns activities) as, for example, the fishing industry in Iceland. The main argument for flexible exchange rates is that the balance of payment difficulties can be avoided and competitiveness of the export industries is ensured. Some of these arguments will be re-examined when the question of appropriate exchange regime for Iceland is discussed.

Inflation in Iceland

A high rate of inflation has been a major characteristic of the Icelandic economy until recently. In figure 6.2 inflation in Iceland is shown as compared to OECD countries. Inflation here is measured as percent changes in the implicit price deflator of the GDP.

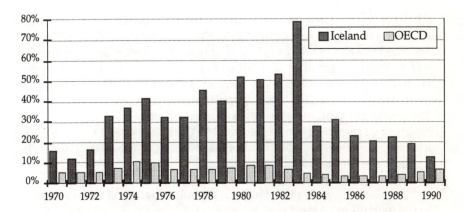

Figure 6.2 **Inflation in Iceland and OECD, 1970–1990. Percentage change in the rate of inflation on the previous year**

Source: Þjóðhagsstofnun (1993).

The average inflation rate in Iceland has been around 36% since 1970 but around 7% in the OECD countries. In figure 6.3 yearly changes in the consumer price index from 1945 and the supply of money (M3) minus the growth rate of GDP from 1961 is shown.

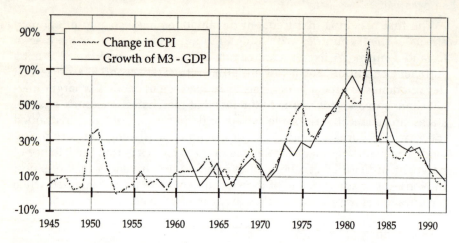

Figure 6.3   **Inflation and the supply of money in Iceland, 1945–1992. Consumer price index and excess growth of money supply over GDP, percentage change on the previous year**

Sources: Þjóðhagsstofnun (1993) and Seðlabanki Íslands, direct communication.

Changes in the consumer price index can be regarded as a measure of the price level and the rate of inflation. It is a subject of debate whether the difference between M3 and GNP is simply best regarded as another measure of inflation or whether such an increase in money supply in excess of GDP actually causes inflation. The figure shows a remarkably strong correlation between these two seemingly different quantities.

The problem of inflation has been of major interest in Iceland for a long time. Most governments since the World War II have been occupied with the solution of this problem. Two closely related questions concerning inflation are of interest here:

1   How do relative prices change, especially wages and interest rates, if the general price level rises?

2   How does a change in the price of certain important commodities such as oil lead to an increase in the general price level?

If relative prices are not changed during times of inflation the effects of inflation are not significant. This situation would be equivalent to a situation where all prices were index-linked. Even if all actors adjusted rapidly to the price level, prices could not adjust instantly in times of accelerating inflation. Therefore it would be possible to transfer values through the money supply or change relative prices for the time being by creating accelerating inflation.

Real interest rates became grossly negative when inflation started to rise in the seventies, as shown in figure 6.4. The value of at least one 'commodity' was diminished greatly, namely the value of monetary savings. In that process real resources were transferred from lenders to borrowers. This encouraged unprofitable investment which slowed down economic growth.

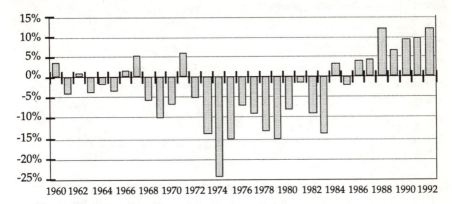

Figure 6.4   Average annual real lending rates in Iceland, 1960–1992
Source: Þjóðhagsstofnun (1993).

General indexing of loans and deposits was allowed in 1979. This measure was initiated as an attempt to increase monetary savings that had declined rapidly in the seventies and to protect the real value of loan capital.[4] In 1983 34% of loans from commercial banks were already indexed and 29% of deposits (Hörður Sverrisson, 1984:71).

In Iceland wage indexation has been on and off since 1939. Indexation of wages was abolished from 1961–1964 and again from 1983 onwards. By indexing wages the possibility to lower real wages by increasing the money supply is reduced. In figure 6.5 changes in the price level and changes in earnings indicate that earnings have generally followed the inflation rate with a time lag except in 1983. This points to real wages not having been greatly affected by inflation (earnings are used as a proxy for real wages).[5]

High rate of economic growth during the periods of high inflation has also helped to reduce the negative effects of inflation on real wages. Unemployment was virtually non-existent in Iceland between 1970 and 1990 or less than 1% most years (Þjóðhagsstofnun, 1993:98–99). The connection between inflation and unemployment ('Philips curve') can best be explained if it is assumed that higher inflation leads to lowering of real wages. If there is no lowering of real wages or no money illusion then increased inflation would not reduce unemployment unless for a short time namely if inflation was accelerating. In 1989 unemployment started to increase but inflation to decline as economic growth stopped. Unemployment was up to 3.0% in 1992 and 4.3% in 1993 while the inflation rate was down to 4.1% in 1993 (Seðlabanki Íslands, 1994:30–31).

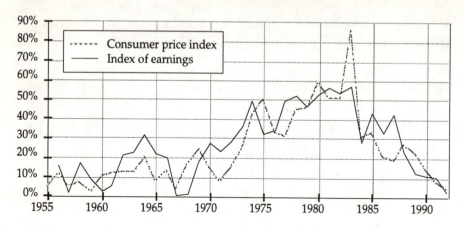

Figure 6.5   **Inflation and earnings in Iceland, 1956–1992. Index of consumer price and earnings, percentage change on the previous year**

Source: Þjóðhagsstofnun (1993).

Under a system of general indexation of wages it is easy to see how a change in the price of a commodity that is included in the measure of the index will lead to a rise in wages. This in turn leads to a rise in the price of other commodities and so the general price level increases. The increase in the price level following the rise of a single commodity depends on the weight of the commodity in the index. But changes in relative prices can lead to an inflation although wages are not linked to an index. This happened in 1974–1975 after the rise in the price of oil and again to a lesser extent in 1979.

The fishing fleet in Iceland uses large amounts of oil and its profitability depends to a large extent on the price of oil. In figure 6.6 an index of gasoline prices and the volume of oil used by the fishing fleet is shown. It can be seen that although the price of oil rose sharply in 1973 and again in 1979 the amount used did not decrease as expected. This was doubtless due to a subsidisation of oil prices although this trend also reflected the change from boats to powerful stern-trawlers that occurred at this time in Iceland. In 1974 an 'Oil Fund for Fishing Vessels' was established that subsidised oil prices to the fishing fleet until 1976.[6] In 1979 a temporary oil charge on raw fish prices was legislated to alleviate the cost effects of the rise in oil prices on the fishing fleet.[7] The government was running a relatively large deficit in 1974 and 1975.[8] The subsidisation of oil prices had negative effects on the trade balance and lead to a monetary expansion that increased inflation.[9]

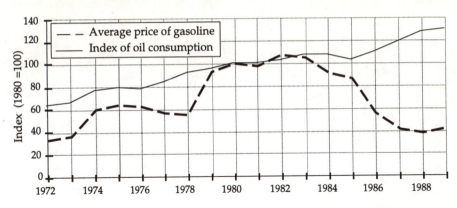

Figure 6.6 **Oil consumption of the fishing fleet and gasoline prices in Iceland, 1972–1989**

Sources: Ragnarsson (1990) and the Ministry of Industry, direct communication.

Exchange rates and inflation

The high rate of inflation has been coupled with frequent devaluations of the Icelandic króna. In figure 6.7 it can be seen that the price of foreign currencies follows the level of prices although the changes in the price of foreign currencies are generally less than in the level of prices.

Since 1989 a policy of fixed exchange rates and low inflation has been pursued in Iceland. The result has been that inflation has been reduced from over 20% in 1989 to 4.1% in 1993. The currency was, however, devaluated in November 1992 after unrest in the international currency markets followed by devaluation in Sweden, Spain and Portugal. Following the announcement of a sharp reduction in cod quota for the fishing year 1993–1994 the króna was devalued again in June 1993 around 6%. This devaluation was done despite wage rates having been stable and the external trade balance steadily improving due to sharply reduced imports. Imports decreased by 7.8% from 1991 to 1992 (Seðlabanki Íslands, 1994:27). The immediate effect of the devaluation was a temporarily sharp rise in interest rates and in the inflation rate. This devaluation shows clearly that price levels are very sensitive to the price of foreign currency and that a very large rise in interest rates is required to counter inflationary effects of a devaluation. It also shows how heavily the interests of the fishing industry weigh in monetary policy in Iceland.

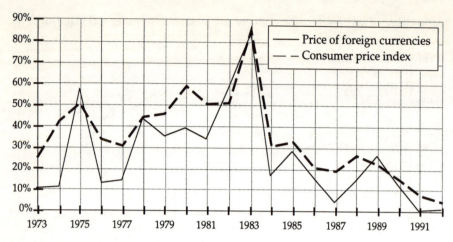

Figure 6.7 Inflation and exchange rates in Iceland, 1973–1992. The consumer price index and the effective price of foreign currencies, percentage change on the previous year

Source: Þjóðhagsstofnun (1993).

Because the economy is heavily dependent on foreign trade there is little doubt that a stable exchange rate is a necessary condition for price stability although it may not be a sufficient condition in the short run. This leads to the question of an appropriate exchange regime for Iceland.

An appropriate exchange regime for Iceland

Paul Krugman (1991) has examined the question whether Iceland should maintain its traditional exchange rate flexibility or join the European Monetary System (EMS) or whether it should seek a compromise solution.

The argument for flexible exchange rates is that it is easier for an economy to adjust to shocks that require a general fall in costs and prices in that economy relative to economies elsewhere. Krugman (1991:3–4) identified four main factors that determine the usefulness of exchange rate adjustment. These are size of external shocks, labour mobility, wage flexibility and wage indexation. A country with diversified exports and a stable world demand will not have as much need for a relative price adjustment as one with a limited number of export industries subject to erratic swings. An economy that can export labour to other nations when faced with an adverse shock will be in less need of an exchange rate adjustment. An economy with flexible wages can substitute wage cuts for devaluation. An economy in which wages are explicitly or implicitly indexed to the exchange rate, so that nominal devaluations lead at best to short-lived real depreciation, will have little to gain from exchange rate flexibility. Krugman considers the effects of the degree of capital mobility to be ambiguous. If a decline in exports is perceived as temporary, demand will remain high and a fall in employment might be limited. If the adverse shock is seen as more sustained the fall in investment might be so large that the current account moves towards surplus.

Krugman (1991:7) then discussed microeconomic advantages of a fixed rate and/or common currencies that arise from lesser transaction costs and lesser uncertainty. These gains depend on the extent of trade and investment relations between the relevant countries. A nation with a high ratio of external trade to GDP will gain more from common currencies. The incentive to join a common currency area will be greater if more of one's trading partners are already part of the area. Direct transaction cost savings that would be achieved by a full currency union are about 3% of trade value. This low cost savings ratio means that the case for a fixed exchange rate must rest either on unclear gains from exchanging the unit of account function of money, or from alleged gains in monetary discipline and credibility. A currency peg in an inflation-prone nation provides monetary discipline which in turn gains the country valuable credibility with both financial markets and wage and price setters.

Krugman then examined the Icelandic situation in terms of key factors relevant to the choice of an exchange regime. On the criterion of export instability he finds that Iceland has a better argument than any other OECD nation. On the criterion of labour mobility he asks whether Iceland is comparable to a fishing district within a larger nation such as Canada's Maritime Provinces. The answer is negative. For such sub-national units flexible exchange rates are not proposed because they are very open to trade and can adjust to adverse shocks partly through out-migration. In the case of Iceland he finds that out-migration during slumps in the economy has been too limited to offer an escape valve and is unlikely to do so in the future because of remoteness from other nations, language differences and a high standard of living. On the criterion of wage-flexibility he says that Iceland's labour market institutions are set up in a way that should make wages more flexible in response to external shocks than those of most other nations. The reasons are that wages in the fishing industry are subject to a revenue-sharing arrangement dependent on catch and world-prices and the labour force is highly unionised and tends to engage in coordinated central bargaining. On wage indexation he points out that indexation has been progressively weakened so that nominal devaluations are now less likely to be passed through into wages as happened before 1979.

Krugman (1991:17–18) observes that Iceland has a smaller ratio of trade to GDP than could be expected from its small size. Other nations such as Ireland have higher ratios. This is explained by a geographical isolation, lack of intra-industry trade and resource based economy.[10] This implies lower micro economic costs of having its own currency than could be expected from the size of the population. Krugman then discussed the inflation problem in Iceland and finds that a commitment to a stable exchange rate or a common currency would discipline the central bank but hardly improve credibility in financial markets and labour and product markets.

Iceland faces two main options for exchange rate policy, either an independent exchange rate policy or to join the European Monetary System, eventually adopting a common European currency if it will be realised. Krugman (1991:23–24) concludes that there are strong advantages combined with an independent exchange rate as a valuable tool of macroeconomic stabilisation if domestic political pressures do not lead to unwarranted depreciations and high inflation. There is no compelling economic case for Iceland to move towards institutional monetary union with European nations. The compelling case, if there is one, is largely political.

Institutional monetary reform

Krugman's paper called on comments from several Icelandic economists. It was pointed out, for example, that the Central Bank had conflicting goals in its statutes which made it difficult for it to pursue an independent policy. The Government could borrow almost at will in the Bank within the fiscal year and the Government also diluted the interest rate policy of the Bank by trying to fix the (real) interest rate of its bonds. More price stability could hardly be obtained without institutional reform (Guðmundur Magnússon, 1991).

To prevent mistakes in monetary policy it seems necessary to transfer the control of monetary matters from the government to an independent professional body namely an independent central bank. The low inflation rate in the OECS states shows the advantages of an independent central bank. The Eastern Caribbean Central Bank conducts monetary policy for the OECS countries according to preset rules. Strict limitations are imposed on the ability of governments to monetise public debt by borrowing and the bank is obliged to maintain a foreign exchange cover equivalent to at least 60% of its demand liabilities. It should be pointed out that the unemployment rate is high in these countries (The World Bank, 1992:406).

A new law concerning the Central Bank of Iceland has been proposed. In this law the Central Bank is given one main objective; to preserve the value of the króna. In an explanatory note to the proposed law it is made clear that the government is the final authority concerning decisions on currency rates (Alþingi, 1992:32). Given the high sensitivity of the price level to import prices it is not possible to see how the Central Bank can pursue its goal of preserving the value of the króna without a complete control over the exchange rate. Therefore it is necessary to revise the proposed law and make the Central Bank fully independent from short term political influences.[11]

In a report to the prime minister's office on economic policy in Iceland the following points were made *inter alia*:

1   Devaluations have mainly been performed to secure the profitability of the fishing sector and have amounted to a transfer of income from the public at large to the export sector. Devaluations have not been used aggressively to increase market shares abroad.

2   In a fixed exchange rate system monetary policy must be aimed at supporting the exchange rate policy. Monetary policy cannot any more be used independently for other purposes. It would therefore be of help to have access to additional policy instruments such as variable export price equalization payments and/or a variable resource tax in the fisheries.

3   Fluctuations in real wages have until now closely followed those of the real exchange rate. Economic policy measures have often consisted in bringing about changes in real wages by means of inflation in order to gain flexibility in the economy. With a more disciplined exchange rate regime wages will have to be in accordance to it in order to preserve its credibility (Hagfræðistofnun Háskóla Íslands, 1991a:7–8).

Conclusion

National currency is a symbol of independence and the possibility of devaluation gives room for a manoeuvre in certain situations. The Icelandic experience shows that it is politically very difficult to resist devaluation when the main export industry demands it. The consequence is most often high inflation rates. Devaluation cannot be used to increase the market share of supply constrained exports such as marine products. In Iceland some industries such as the tourist industry might expand after devaluation. Against this must be weighed the loss of real income that occurs because foreign employers such as the NATO defence force and the aluminium smelter pay less for services and wages (in foreign currency). The disadvantages of devaluation seem more pronounced than the possible advantages. Fixed exchange rates or pegged currency seem the best choice for monetary policy and stability in small states.

The result of the empirical analysis in the last two chapters indicates that economic characteristics of small states do not imply a serious economic vulnerability at least not if monetary problems are avoided. More theoretical analysis is needed to support this view. The next chapter looks behind the curtains and discusses the basis of trade.

Notes

1   Antigua and Barbuda, Dominica, Grenada, St. Lucia, St. Vincent and the Grenadines, St. Kitts and Nevis.

2   Economic growth in the OECS countries was 4.7% between 1980–1989, 5.3% in 1990 and about 3.0% in 1991. See The World Bank (1992:406–418).

3   The main condition for optimum currency area is that factors of production are mobile within the area but immobile internationally.

4   Total bank deposits as a percentage of GDP fell from 36% in 1970 to 20% in 1980 (Þjóðhagsstofnun, 1993:166).

5   Indexation of wages is not a perfect protection of buying power against inflation. The index is calculated ex post (mostly monthly, usually every 3 months). The time-lag between an index-linked rise in wages and an increase in prices is likely to cause some loss of buying power.

6   The fund was financed by special export levies on fish products (OECD, 1974:37).

7   The charge was initially 2.5% but increased to 7.5% later in the year (Þjóðhagsstofnun, 1980:77).

8   The deficit amounted to 2.7% of GDP in 1974 and 2.8% in 1975 but there was a surplus of 1.0% in 1979 (Þjóðhagsstofnun, 1993:59).

9   The balance of trade was -10.4% of GDP in 1974, -10.1% in 1975 but +1.0% in 1979 due to good fish catches (Þjóðhagsstofnun, 1993:125–126).

10   The meaning and measurement of intra-industry trade is discussed in the next chapter. The general principle is that 'the higher the proportion of a country's export which is matched by imports falling under the same heading of international trade classification the greater is the importance of intra-industry trade' (Frederick V. Meyer, 1978:43).

11   More detailed arguments for the opinions expressed in this paragraph have been presented in Ólafsson (1993a) and Ólafsson (1993b).

7 The theory of international trade as applied to small states

Small states have on the average a high ratio of trade to GDP, a high content of primary products or natural resources in their exports and a high concentration in trade as was shown in chapter 5. There is, however, a wide variation within the group of small states in those characteristics and many larger developing and developed states can also be characterised in this way to some extent. This raises the question whether trade and production of small states can be characterised in a more specific way. It turns out that this is possible by examining the status of small states in the light of the theory of international trade.

In the world of classical trade theory nations gain by trading when a nation exchanges goods, in which production the nation has a comparative advantage, for different goods, in which the nation has a comparative disadvantage. The size of a nation is not of much concern if the comparative advantage is explained within the framework of a factor proportion theory like the Heckscher-Ohlin theory where only differences in the relative abundance of factors of production are regarded as the basis of a comparative advantage. If demand factors are also taken into account it can be shown, for example, that a small state which trades with a large one captures all the gains from the trade because its supply does not depress prices in the large nation ('the importance of being unimportant'). Smallness is here associated with a relatively small output of the traded good. In reality extensive specialisation in the export industries may lead to a strong market position for a small state. Iceland, for example, supplies around 18% of cod consumed in the United States.[1] Iceland's share of world exports in certain fish products was around 12% in 1991.[2]

The classical theory shows how foreign trade based on comparative advantage can help the small state to overcome the limitations of a small

101

domestic market, but the law of comparative advantage (or absolute advantage which is a special case) explains only why nations gain from trade. The basis of trade or the explanation of why some nations possess or obtain comparative advantage is a more complicated problem. Trade can be based on factor endowment, increasing returns, demand conditions and technological gap for example. Here it will be established that international trade of small states is characterised by inter-industry trade rather than intra-industry trade. This implies that the most important basis of trade for small states is a relative difference in factor prices or factor endowment in accordance with the Heckscher-Ohlin theory of trade.

Although it can be shown that conditions for intra-industry trade based on increasing returns activities are lacking in small states this does not mean that small states can not develop and compete internationally in technically advanced industries. Examination of product cycle and technological gap theories cast light on this important possibility.

The small economy may face disadvantages in foreign trade that reduce the benefits. These disadvantages are both of a political and economical nature and arise mainly because of the lack of diversification in production and exports and a lack of power in international relations. The competitive forces affecting small states in the world economy and their ability to attract investment will be the subject of the next chapter.

The development of international trade

A notable feature of the international trade in the first decades of the century was that manufactures and primary products constituted an almost constant ratio of world trade around 40% and 60% respectively (W. Arthur Lewis, 1952). In addition the growth rate of trade in primary products correlated closely with the growth rate of manufacturing production. It was probably because manufactures were mainly exported in order to obtain the means to buy raw materials and/or food.

Although industrial countries were both primary traders and producers, the countries of recent settlement colonies and peripheral countries of Europe were the main suppliers of primary products in exchange for industrial goods. After World War II trade in manufactures especially between industrial states began to grow rapidly and faster than production of manufacturing (tenfold increase in trade against close to a fourfold increase in production 1948 to 1979). Trade in primary products also grew but on the average at a similar rate as the production of manufactures. The former 40% share of manufactures in world foreign trade increased to around 60%. This rapid increase in the trade of manufactures between industrialised countries followed the growth of research and development

102

intensive industries. Manufacturing firms now had to sell the largest possible amount in the shortest possible time in order to recoup their research and development costs before their products became obsolete. Access to markets for outputs became even more important than cheap supplies of inputs. In addition to inter-industry trade and specialisation which characterised the international system before World War II, there came intra-industry trade and specialisation (Meyer, 1979).

Intra-industry trade arises when countries exchange similar goods or, in statistical analysis, goods that fall under the same heading in tariff schedules. France, for example, produces cars and exports some of them to Germany while Germany also produces cars and sells some of them in France. The French cars (in the same price range) are functionally almost equivalent to the German cars, the choice between them is based on taste and quality rather than great differences in size or performance. The same applies to many home appliances such as washing machines and radios etc. A great deal of trade is based on such differentiated products rather than on products that are completely different as when Iceland exchanges fish for cars. Intra-industry trade is not based on factor abundance in the same way as inter-industry trade as will be explained later.

The basis of inter-industry trade

Adam Smith in his book *The Wealth of Nations*, published in 1776, stated that free trade can be beneficial to nations when based on an absolute advantage. Each nation could specialise in the production of those commodities in which it was more efficient than other nations and import those commodities which it could produce less efficiently. Adam Smith was arguing against the mercantilist orthodoxy which maintained that the way for a nation to become rich and powerful was to export more than it imported. Ricardo strengthened the case for free trade when he in 1817 published his book *Principles of Political Economy and Taxation*, in which he presented the law of comparative advantage. He demonstrated by means of numerical examples that even if a nation is less efficient than another nation in the production of (both) commodities there is still a basis for mutually beneficial trade. The first nation should specialise in the production and export of the commodity in which its absolute disadvantage is smaller and import the commodity in which its absolute disadvantage is greater.[3]

In the Ricardian model only one factor of production, namely labour, is allowed so international trade is based on differences in productivity of labour or technological differences between nations. The Ricardian model has been tested by comparing labour productivity and export shares in the

US and UK. The results have consistently shown positive and significant correlation between these two variables (Alan V. Deardorff, 1984:477). Salvatore (1990:142) concludes that although the simple Ricardian trade model has been empirically verified to a large extent it is unsatisfactory because it assumes rather than explains comparative advantage.

The Heckscher-Ohlin theory of trade

Bertil Ohlin (1967) using some ideas from his teacher Eli Heckscher explained differences in (pretrade) prices of goods between regions as resulting from differences in factor endowment. He criticised the classical comparative cost theory of international trade, namely the Ricardian model, whose assumptions of only one variable factor of production he finds too restrictive. Ohlin gives a simplified theory of trade between regions when the factors of production are immobile between regions but freely mobile within regions. Later he discusses in detail the difference between regions and nations, the implication of factor mobility, commodity movements, obstacles to trade and locational aspects of trade. It is mainly the first part of his theory or the simplified view of trade which has been used in models of international trade.

Ohlin(1967:7) assumes that regions are differently endowed with facilities for the production of various articles. This creates a condition for trade:

> Each region is best equipped to produce the goods that require large proportions of the factors relatively abundant there; it is least fit to produce goods requiring large proportions of factors existing within its borders in small quantities or not at all. Clearly, this is a cause of interregional trade, just as varying individual ability is a cause of individual exchange.

The immediate cause of trade is always that goods can be bought cheaper from outside in terms of money than they can be produced at home and *vice versa*. The price mechanism rests on four basic elements. Demand is based on the wants and desires of consumers and the conditions of ownership of production. Supply of goods depends upon the supply of productive factors and the physical conditions of production. The latter determine the combinations of factors or the technical process. When the relation between these elements is different the relative commodity prices are also different. A large supply of a factor makes it comparatively cheap unless in the unlikely event that inequality of demand balances this inequality of supply. Each region has an advantage in the production of commodities into which enter considerable amounts of factors abundant and cheap in that region.

It is the relative element in this analysis of Ohlin which is important. Every small state must have a comparative advantage in some products even though total supply is limited, so trade (and specialisation) is always an improvement on autarky.

Ohlin (1967:19, 37) states two conditions of trade. The first condition is that some goods can be produced more cheaply in one region than in another. In each region the exported goods contain relatively great quantities of the factors cheaper there than in the other regions, whereas the goods that can be more cheaply produced in other regions are imported. The second condition of trade is based on the effects on interregional prices that a lack of internal mobility and lack of divisibility of productive factors generates. Not everywhere are commodity prices equal to costs of production because of several factors called 'economic friction'. Producers may sell at different prices in different markets and risks differ between regions when production factors are directed towards certain uses. Also a lack of divisibility makes production on a large scale more efficient up to a certain point than production in small quantities. The economies of large-scale production make interregional division of labour profitable, irrespective of differences in the prices of the factors of production. The advantages of specialisation resulting from large-scale production encourage interregional trade.

This second condition of trade is important in the case of small states although trade of small states is not based on large-scale production. The benefits of large-scale production at the plant level cannot be obtained in a small state unless the product is exported because of a limited home market. Indivisibilities in production are more pronounced as the state gets smaller especially indivisibilities due to a lack of auxiliary services and supporting industries.

Ohlin finds that the relative scarcity of the productive factors or factor prices are reduced when two regions trade. But there is not a complete equalisation:

> This tendency towards an equalization of both factor and commodity prices is the natural consequence of the fact that *trade allows industrial activity to adapt itself locally to the available factors of production*. Industries requiring a large proportion of certain factors gravitate toward regions where those factors are to be found in large quantities and therefore at low prices. Producers, no less than consumers, look for the cheapest market. In brief, an uneven distribution of productive factors will, unless it is balanced by a corresponding geographical unevenness of demand, cause factor prices to be different in the various regions, and bring about a certain division of labor and trade between them. (Ohlin, 1967:26)

The complete equality of factor prices is impossible when the costs of transport and other impediments to trade are taken into account. Ohlin (1967:40) concludes that interregional trade has a tendency to reduce the disadvantages of both the lack of mobility and the lack of divisibility of the factors of production and can be regarded as a substitute for geographical mobility of the productive factors.

This theory explains how international trade provides an escape mechanism from suboptimal economy and the negative effect of the area factor (scattered settlement structure) in small island states where natural resources such as fishing grounds or natural beauty constitute immobile factors of production. A technical progress which lowers transportation and communication costs serves to strengthen the competitive position of small island states although factor prices will not be equalised.

Ohlin modifies his theory and applies the interregional trade theory to international trade. The different factors of production labour, natural resources and capital are discussed and defined more fully. He states that natural resources can be regarded as independent factors of production:

> Although many of their differences are from an economic standpoint unimportant, a sufficient number of economically essential inequalities remains to necessitate a division of natural resources into a great number of factors. Such a procedure does not present the same kind of difficulties as does the subdivision of labor into several separate factors. Whereas transition from one labor group to another is in many cases comparatively easy and materially affects the supply within each group, such a transition between different factors of nature is possible only by means of capital investment. (Ohlin, 1967:54)

Most small states export resource intensive products or services such as tourism which is based directly and indirectly on natural resources. Capital and labour can often be substituted for each other but natural resources are not easily substituted for each other or for capital and labour except, as Ohlin points out, by means of capital investment. A comparative advantage of small states would therefore be difficult to explain if natural resources such as rich fishing grounds, sunny beaches, oil or copra are not counted as separate factors of production.

Ohlin discusses more fully the implications of the different quality of factors and commodities for international trade and specialisation. The tendency towards factor price equalisation is qualified. 'Trade does not tend to equalize factor prices when quite different factors are closely competitive, being used in one industry to produce the same or similar commodity while otherwise rendering quite different services. This is by no means rare, on the contrary, a great many goods are produced by means of widely different technical processes' (Ohlin, 1967:69). Ohlin continues to refine and add to his theory and takes into account the terms of trade, locational elements and monetary factors.

The Heckscher-Ohlin theory as a formal Model

The Heckscher-Ohlin theory is usually represented formally as a simple model based on the following assumptions (Salvatore, 1990:104):

1   There are two nations, two commodities (X and Y) and two factors of production.

2   Both nations use the same technology in production.

3   Commodity X is labour intensive and commodity Y is capital intensive in both nations.

4   Constant returns to scale in the production of both commodities in both nation.

5   Incomplete specialisation in production in both nations.

6   Equal tastes in both nations.

7   Perfect competition in both commodities and factor markets in each nations.

8   Perfect factor mobility within each nation but no international factor mobility.

9   *No transportation costs, tariffs, or other obstructions to the free flow of international trade.*

Some of these assumptions were made only implicitly by Heckscher and Ohlin. From these assumptions several related theorems have been deduced. The most important are (R. W. Jones and J. P. Neary, 1984:15):

1   *Factor-price equalization theorem.* In its global form, this theorem states that, under certain conditions, free trade in final goods alone brings about complete international equalisation of factor prices. In its local form, the theorem asserts that, at constant commodity prices, a small change in a country's factor endowments does not affect factor prices.

2   *Stolper-Samuelsson theorem*. An increase in the relative price of one commodity raises the relative return of the factor used intensively in producing that commodity and lowers the real return of the other factor.

3   *Rybczynski theorem*. If commodity prices are held fixed, an increase in the endowment of one factor causes a more than proportionate increase in the output of the commodity that uses that factor relatively intensively and an absolute decline in the output of the other commodity.

4   *Heckscher-Ohlin theorem*.   A country has a production bias towards, and hence tends to export, the commodity that uses intensively the factor with which it is relatively well endowed.

The 'proofs' mainly by geometrical means are supplied in standard text-books in international trade theory. The Heckscher-Ohlin theorem is the basic theorem and the factor equalisation theorem is deduced from it. The other two theorems show the magnification effect of factor growth or price increases.

Criticism and tests of the Heckscher-Ohlin theory

John Ford (1963) has argued that two theories of trade emerge from Ohlin's work, basic and subsidiary. He states that Ohlin ignored the postulates of the subsidiary model and regarded them as unimportant in reality. The subsidiary model discusses what the basis of trade would be on more real-istic assumptions and has more importance than has been attached to it so far (Ford, 1963:459).

It can hardly be accepted that Ohlin ignored the postulates of the sub-sidiary model. He presents the 'basic' model in the first part of the book but devotes most of his effort to make the analysis more realistic. It seems that a greater attention has been paid to the 'basic' model by *other* economists because this basic model can be fitted into the formal apparatus of modern economics.

Ford discusses what makes a 'good' theory and states that the first and perhaps most vital criterion of a 'good' theory is that its premises must be realistic. The second criterion is that the conclusions drawn from such premises must be logically correct.

From our discussion of this theory it will be evident that as 'realism' is to be regarded as the sternest test for a theory, the Ohlin-Heckshcer analysis must be held to be an unsatisfactory piece of

108

theorising, the assumptions of the basic Ohlin-Heckshcer model are nearly all mere abstractions: they are really contrary to what we observe around us. So that the Ohlin-Heckshcer premises of perfect competition, no differences in the quality of factors, no product differentiation, lack of increasing returns to scale industries, fixed production functions are all unrealistic. (Ford, 1963:469–470)

Ford (1963:470) finds no faults with the logical aspect of the theory, there are no flaws in the conclusion that Ohlin and Heckscher reach from their premises.

In this discussion Ford contradicts clearly the positivist view of Milton Friedman, for example, in regard to the criterion of a good theory. Friedman (1968:516–517) states that the realism of the assumptions of a theory is irrelevant, the test of a theory is its applicability and predictive power. There are many examples of correct predictions from unrealistic premises and Ford's views regarding the criteria of a good theory are therefore debatable. Friedman's views are also debatable but, leaving the philosophical point aside, the question here is how successful Ohlin has been in simplifying reality in his assumptions and how important the individual assumptions are for the theory. By relaxing some of them different bases of trade are possible as will be seen in the next section. In the case of small states the assumption of a lack of increasing returns to scale seems to hold, they are generally price takers in foreign trade and standardised products are important in their export trade (oil, copra, fishmeal). The assumption of fixed production functions and quality differences of labour are probably not damaging to the theory in the context of small states.

The Heckscher-Ohlin theorem has been subject to several empirical tests. The result so far seems to be inconclusive. The first and best known is a test conducted by Leontief published in 1953. He computed the ratio of capital stock to the number of workers in export industries and import-competing industries for the United States in 1947. His result was that the US was exporting labour-intensive goods in exchange for capital-intensive imports. This result was called the Leontief paradox because it was generally assumed that the US was capital abundant and labour scarce relative to the rest of the world. Charles P. Kindleberger and Peter H. Lindert (1978:98) are of the opinion that differences in skills and human capital between nations and differences in natural resource endowment are the most important sources of the Leontief paradox. Capital is not the most abundant factor in the US relative to the rest of the world. The United States is most abundantly endowed with skilled labour and farmland they say.

Salvatore (1990:159–160) concludes that the Heckscher-Ohlin model offers a reasonable explanation of international trade in non-industrial goods, especially trade based on the natural resource endowments of the nation, but leaves a substantial portion of international trade unexplained especially trade in differentiated manufactured products among industrial countries. Theories based on economies of scale, differentiated products and changes in technology are required to account for such a trade.

Iceland's comparative advantage

According to the Heckscher-Ohlin theory Iceland should export the commodities in the production of which a great deal of its relatively abundant and cheap factors are used. Iceland is blessed with rich fishing grounds and abundance of geothermal energy and hydropower. Foreign trade in Iceland should therefore be based on exports of fish products and energy intensive industries. This turns out to be the case. The fishing industry has been the most important export industry in the Icelandic economy. Marine products are over 70% of merchandise exports and over 50% of total export earnings. The energy resource has also been used in export production.

Iceland has comparative advantage or rather absolute advantage in fisheries. This is confirmed by looking at value added per fisherman in Iceland compared to other nations. In Iceland the value added of the catch by approximately 7,100 fishermen was 476 million ECU in 1990, in Denmark 7,300 fishermen obtained 336 million ECU and the total for the EU was 297,000 fishermen and value added of 4,632 million ECU. An Icelandic fisherman produced 4 times more valued added than the average fisherman in the EU.[4] The fishing industry is heavily subsidised in most neighbouring countries but only marginally in Iceland.[5]

Catches and market prices are unstable so changes in economic growth are often closely related to the fortune of this industry. When Iceland established a 50 mile fisheries zone in 1972 and subsequently extended it to 200 miles in 1975, investment in the fishing sector was increased to be better able to utilise the fishing stocks. Many new stern-trawlers were obtained and new fish-processing plants built or rebuilt. Fish catch has, however, not increased as much as the capacity of the fishing fleet or the processing plants. The most important fishing stock, namely cod, has weakened, probably due to overfishing. Quotas on total allowable catch, first applied systematically in 1984, had to be reduced year after year.[6]

In figure 7.1 the catch of cod, herring and capelin is shown and the total catch of all demersal species in cod equivalent units.[7]

110

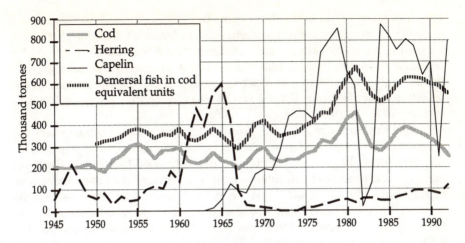

Figure 7.1   Fish catch in Icelandic waters, 1945–1992
Source: Þjóðhagsstofnun, direct communication.

Growth of exports has slowed down or halted as shown in figure 7.2 where indices of export value of marine products, manpower and fixed assets in fishing and fish production are shown.

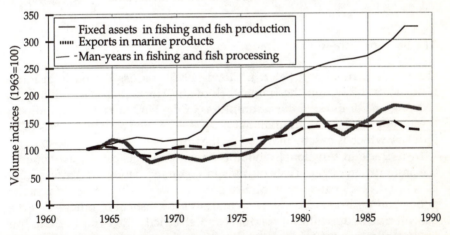

Figure 7.2   Expansion in the fishing sector in Iceland, 1963–1989. Fixed assets, export value and man-years in the fishing sector
Source: Þjóðhagsstofnun, direct communication.

It seems that the fishing sector is heading towards sharply decreasing returns on investment. Economic growth based on volume growth in the fishing sector cannot be assumed in the future which means that Iceland must put more effort into other export industries or services. Despite the problems facing the fishing industry it is clear that Iceland still has a marked comparative advantage in fisheries. Profits could be increased by reducing overcapacity in the fishing fleet by a resource tax or auctioning of fishing rights. Overfishing and overcapacity is a global problem which has resulted in depletion of fishing stocks. It is likely that prices of fish will increase in the future.

In addition to the rich fishing grounds Iceland has abundance of energy resources in the form of hydro-electric power and geothermal power. This resource has only been partly utilised. It is estimated that 28,000 GWh per year can be produced relatively cheaply compared to other sources of energy but only about 5,000 GWh had been utilised in 1993 (Seðlabanki Íslands, 1994:33). Much less has been utilised of the geothermal energy. Development of energy intensive industries that utilise hydro-electric power requires more capital than is available domestically. Foreign investment is therefore necessary in this sector. There are plans to build a new aluminium smelter in Iceland but depressed markets for aluminium have prevented action so far (another interesting possibility is the production of magnesium metal from sea-water). The Icelandic experience shows that an abundant natural resource such as hydro-energy is not a sufficient basis for trade and development if other necessary factors such as capital are scarce or demand conditions are unfavourable.

The basis of intra-industry trade

Herbert G. Grubel and Peter. J. Lloyd (1975) discuss the basis of intra-industry trade within the framework of the Heckscher-Ohlin theory by relaxing one or more of the assumptions of the H-O model. Intra-industry trade is defined for computational purposes as the value of exports of an industry which is exactly matched by the imports of the same industry.

Intra-trade in functionally homogeneous products can be explained if the assumption of zero costs of transport, storage, selling and information is relaxed. This category contains border trade in bulky goods, heavy goods and perishable products, in addition to seasonal or otherwise time-based trade in agricultural products and electricity. Re-export or entrepot trade leads to intra-trade as well as trade in services such as financial, insurance, shipping, brokerage and related services that are purchased by the exporters and importers of goods (Grubel and Lloyd, 1975:72–84). Intra-trade of this type is known in small states. Iceland, for example, imports shrimp and cod caught by Russian trawlers for further processing and re-export. In this way effects on production and employment caused by seasonal variations in fish catch in Icelandic waters can be reduced.[8]

Grubel and Lloyd (1975:85–101) attempt to explain the trade in close-substitute manufactures produced under increasing returns to scale by relaxing the H-O assumption of linearly homogeneous production functions and homogeneity of commodities with respect to all functional characteristics and with respect to location, time of use and packaging. Economies of scale are a function of the length of production runs of each product, due to reduced downtime of machines, greater specialisation of machines and labour and to smaller inventories. They conclude that economies of scale is quantitatively the most important explanation of intra-industry trade.

Relaxing the assumption that production functions are identical in all countries, Grubel and Lloyd (1975:102–117) try to integrate intra-industry trade with theories of technological gap and product cycle trade. This will be discussed in the next section.

Meyer (1978:214) shows that freer trade is one condition of intra-trade. But other conditions must be fulfilled. First, a short market life of the product relative to the length of the R&D stage, implying production under increasing returns to scale. Second, increased specialisation between producing units within each product range, which implies that trade can grow faster than production, and that such trade reduces and the risk of supply restraint. Third, the use and the user of the product can be identified at the point of production which implies production of specific goods. High research and development cost coupled with a short market life of the product require a lot of capital which is tied up for a long time before there can be a return to it. If R&D costs are to be born by the firm it requires financial strength which is only found in large enterprises. 'The evidence points to that intermediate export concentration, high research intensity and high degree of intra trade went together' (Meyer, 1979:330).

Meyer (1979:336–338) further argues that the main reason for the absence of intra-trade in primary commodities cannot be the lack of access to markets nor the lack of costly R&D. International trade in raw materials was relatively free even in the most protectionism days and no intra-trades emerged. Some primary producing activities are amongst the most R&D intensive ones. In manufacturing, R&D led to improved quality and increased variety of products. It led to specialisation within industries so that the individual producing units produced increasingly different goods. In primary production, R&D led to increased quantity and standardisation. Standardisation in primary commodities means mainly more constant quality and that the individual producing units produce increasingly similar goods. Primary production also takes place under conditions of decreasing returns to scale. One main reason is that the market life of primary commodities is for all practical purposes infinite relative to the R&D stage. The advantage of wider markets can only be used when production increases. When production increases primary producers' unit costs rise. Such producers may therefore prefer protection. From their point of view this gives them additional time in which to sell in an assured market. They get a price above marginal costs and they produce a smaller output at low unit costs. Meyer (1979:338) concludes that 'Today's free trades are intra-trades. The primary producers are not, or not yet, amongst them'.

This statement holds if the existing (domestic) market is large enough to absorb the output before decreasing returns set in. Although the domestic market is too limited in small states for most manufacturing activities, the domestic market is of a certain importance especially to services and some agricultural products such as dairy products where economies of scale are not important or are exhausted in relatively small factories. In Iceland, for example, the agricultural sector depends heavily on the domestic market. Imports of meat and dairy products are not allowed and this has led to high prices for the Icelandic consumers. The recent GATT agreement will gradually open the market for imports, therefore Icelandic agriculture must adjust to increased competition from abroad, but at the same time foreign markets will open up for special Icelandic products such as lamb meat.

In recent years economists have employed new models of trade that allow intra-industry trade without differences in factor proportions. In the monopolistic competitive model each firm is assumed to be able to differentiate its product from other rival firms and prices charged by other rival firms are taken as given. It has been shown how trade in this model leads to intra-industry trade, based on economies of scale, between countries with the same factor endowment (Krugman, 1990:74–78).

David Greenaway and Chris Milner (1986:15) criticise the monopolistic competitive models (the neo-Chamberlinian models) on the grounds *inter alia* that there is no role for demand in determining product variety, the process of adjustment to trade expansion appears to be costless and unless initial factor endowments differ between countries the direction of trade is indeterminate. Krugman (1990:78) seems to think that the conclusions of the monopolistic model are not very sensitive to its assumptions. 'The importance of increasing returns in trade does not stand or fall on the validity of particular interpretations of product differentiation or of twoway trade within statistical classifications.'

The monopolistic model shows explicitly how intra-trade can arise in differentiated products and how the size of the market and trade between similar countries enables economies of scale. This model does not explain how increasing returns occur in production and it is not easy to see how this line of thinking can explain a faster rate of growth in trade than growth of output (of manufactures).

Salvatore (1990:152) makes the following observations about intra-industry trade:

First, while trade in the H-O model is based on different factor endowments (labour, capital, natural resources and technology), intra-industry trade is likely to be larger among economies of similar size and factor proportions.

Second, product differentiation and economies of scale are closely related [...]. International competition forces firms in each industrial nation to produce only a few varieties and styles of a product in order to take advantage of economies of scale and lower per unit production costs. The nation then imports other varieties and styles of the commodity from other nations. Thus inter-industry trade can be explained by the standard H-O model (i.e., by comparative advantage), while intra-industry trade can be explained by product differentiation and economies of scale.

Thirdly, with differentiated products produced under economies of scale, pretrade relative commodity prices may no longer accurately predict the pattern of trade. Specifically, a large country may produce a commodity at lower cost than a smaller country in the absence of trade (because of larger national economies of scale). With trade, however, all countries can take advantage of economies of scale to the same extent, and the smaller country could conceivably undersell the larger nation in the same commodity.

Finally, as contrasted to the H-O model, which predicts the trade will lower the return of the nation's scarce factor, with intra-industry trade based on economies of scale it is possible for all factors to gain. This may explain why the formation of the European Common Market and the great post-war trade liberalization in manufactured goods met little resistance by interest groups.

Intra-industry trade and small states

In a country with a total labour force of two or three hundred thousand at most it is not possible to establish more than a few large companies. The conditions for increasing return activities based on product differentiation, large scale production and high R&D content do not exist in a limited domestic market. Such activities must therefore be export oriented. But even if an export oriented increasing returns industry could be operated an important basis of intra-trade is lacking. Intra-trade implies reciprocity, that is similar markets and structure of the economies involved (except some intra-trade, for example in seasonal products, not based on increasing returns). Due to a limitation of the domestic market the small economy will

115

not be able to absorb the imports just as it can not supply a market for its own industries. This does not mean that advanced technological industries could not be operated in a small state, especially if they are favourably situated with regard to communication and transport costs. Such industries are then likely to be operated in connection with a larger foreign enterprise (as a subsidiary of a multinational company for example).

An analysis was undertaken to examine intra-industry trade in small states. The result is shown in table 7.1 which includes the ratio of intra-industry trade in small states for which data is available up to the year 1989. The index (Grubel-Lloyd index) is calculated at a 2 digit level of SITC which tends to show a higher value of the index than a calculation based on a lower level of aggregation.[9] This table shows that average intra-trade in manufacturing industries in small states is 6.6% when a weighted index is used. For all merchandise exports the result is 16.1%. The only state with high level of intra-trade in manufactures is Malta. This can be explained by a high level of imports and exports in electrical machinery, such as transistors (SITC 7293), probably due to assembly operations in Malta's Export Processing Zone. Netherlands Antilles, Barbados, Bahrain and Bahamas have a high index mainly due to oil refining based on imports and re-exports of oil products. Average level of intra-trade in manufactures was 59% in industrial countries in the year 1978. The extent of intra-industry trade does not seem to be strongly related to country size in the case of industrial countries (Greenaway and Milner, 1986:96, 103). The analysis shown in table 7.1 confirms that the class of small states is characterised by low level of intra-industry trade and as a consequence a high level of inter-industry trade. If small states are confined to the status of inter-traders whether in primary products or not, this will have consequences for their economic prospects and policy. First, the stability of income from trade is likely to be less than in the case of larger states with substantial intra-trade, especially if the small state relies on 'single crop' primary production. Secondly the gains from regional integration will be more limited for small states than larger industrialised states where intra-industry trade is more prevalent as will be discussed in chapter 9. A lack of intra-industry trade opportunities does not prevent the possibility to operate sophisticated small scale industries or services in small states if human capital is available. Increased processing and production for specific buyers is possible leading to greater stability in market prices and national income.

Table 7.1
Intra-industry trade in small states

Country (year)	All industries	Manufactures
Bahamas (1988)	64.4%	5.6%
Bahrain (1988)	47.0%	4.6%
Barbados (1988)	35.4%	23.6%
Belize (1986)	32.2%	17.6%
Bermuda (1985)	10.7%	1.0%
Brunei Darussalam (1986)	0.3%	0.3%
Cape Verde (1985)	3.5%	2.0%
Cayman Islands (1989)	8.8%	6.8%
Cook Islands (1983)	6.5%	6.5%
Cyprus (1989)	16.1%	10.2%
Dominica (1989)	3.1%	2.6%
Faeroe Islands (1989)	13.8%	6.6%
Fiji (1987)	26.3%	8.2%
French Guiana (1989)	7.5%	2.9%
French Polynesia (1983)	8.9%	6.7%
Gabon (1983)	4.7%	3.2%
Greenland (1989)	2.6%	0.5%
Grenada (1986)	7.8%	3.9%
Guadeloupe (1989)	10.5%	2.7%
Guinea-Bissau (1980)	5.8%	2.6%
Guyana (1979)	9.9%	4.6%
Iceland (1989)	11.1%	8.5%
Kiribati (1987)	2.2%	0.0%
Malta (1989)	57.6%	54.0%
Martinique (1989)	16.3%	4.4%
Mauritius (1987)	8.8%	7.5%
Netherlands Antilles (1984)	88.8%	0.4%
New Caledonia (1983)	9.9%	9.5%
Niue (1983)	0.8%	0.0%
Reunion (1989)	4.2%	2.4%
Saint Lucia (1986)	25.6%	22.9%
Saint Vincent and the Grenadines (1980)	13.3%	6.7%
Samoa (1983)	8.5%	4.1%
Seychelles (1987)	28.1%	3.5%
Solomon Islands (1984)	2.0%	0.8%
St. Pierre and Miquelon (1984)	0.7%	0.0%
Tonga (1985)	8.9%	3.5%
Tuvalu (1983)	1.0%	1.0%
Average	16.1%	6.6%

Source: Calculated from United Nations (1991).[10]

Technological gap and product cycles

International trade based on a technological change or a technological progress has been analysed with the aid of two models: The Technological Gap model and the Product Cycle model. In the Technological Gap model trade is based on the introduction of new products and new production processes. The invention or innovation gives the firm or the nation a temporary monopoly in the world market or absolute advantage for a time. But after some time other nations imitate the new technology and the technological gap will be closed (production functions become similar again as assumed in the H-O model). This idea has been generalised into the Product Cycle model. An example of this line of thinking is the work of Seev Hirsch (1967) who divided the production cycle of new products into three phases. The introductory or new phase, the growth phase and the maturity phase. A product is new if it is manufactured by methods which were not previously used in the industry and is based on a recent invention or unfamiliar developments.

The introductory phase of the product cycle is characterised by high unit costs and a labour-intensive production function. Product runs are short and frequent changes and modifications are made. New products contain a high proportion of scientific and engineering inputs. Professional knowledge and experience are most critical for their successful development and their ability to survive the introductory phase. Products that survive the first phase enter the second or the growth phase. Then mass production and mass distribution are introduced. Production runs are lengthened and the production process becomes more capital intensive. An increasing number of firms are attracted to the industry. Entry is technically possible when patents expire and close substitutes develop. Management is especially important in this stage. In the third or mature phase sales volumes level off and the product becomes more standardised. The manufacturing process becomes even more capital-intensive and the optimal size of the manufacturing unit becomes larger yet. Economies of scale become an important factor in determining the comparative strength of individual manufactures (Hirsch, 1967:18–20).

The problem with this theory is how to apply it to products that are being continuously improved without increased standardisation. Many manufacturing industries such as the car industry and also service industries such as insurance and banking, improve their products or services continuously. Such industries pass the first and second stage but do not enter the third phase as the products are differentiated rather than standardised.

Hirsch also discusses the relationship between a country's stage of development and factor endowment and its ability to manufacture competitively products at each stage of the product cycle. He finds that mature products provide an opportunity for developing countries. The industrial leaders such as the United States have the biggest advantage in the growth products but in the early phase the small states have an advantage. He suggests that smaller countries may possess a comparative advantage in certain categories of new products, provided they have at their disposal an adequate pool of comparatively inexpensive scientific and engineering know-how. The industrial leaders are likely to have a competitive edge in products whose development and manufacture involve extensive utilisation of external economies, capital and managerial inputs. The small developed economies in turn might possess a competitive potential in products less dependent on external economies and more dependent on extensive utilisation of scientific and engineering inputs (Hirsch, 1967:34).

Primary products and/or products of a low technological content characterise the export trade of small states at present. This need not be so. Primary production may in this case partly be a characteristic of low economic development. The success of many small firms in all kinds of economic activity suggests that a very small economy could support a number of firms in technologically advanced industries. The prerequisite is substantial investment in education or human capital.

Advanced manufacturing industries exist in states such as Switzerland, Denmark and Sweden which are often considered 'small'. In the case of Iceland there are even a few examples of new export products that have been developed in connection with the fishing industry. The greatest opportunity for small states is probably in computer related products and software industries and similar activities that require investment in human capital but are not dependent on increasing returns or the conditions for intra-industry trade. Activities such as software development can be located in the most remote places if modern communications are provided. In Iceland the use of computers started early and is widespread. A promising software industry has been developing during the last decade.[11] Financial software, for example, had to be adapted to Icelandic laws and regulations and to the Icelandic language. The sophisticated demand at home has lead to the creation of software houses which are now selling advanced products in foreign markets.[12]

Conclusion

The Heckscher-Ohlin theory is sufficient to explain the characteristics of foreign trade of small states. Most small states depend on inter-industry trade which is based on the export of standardised products or food

119

usually with a high content of natural resources. Specialisation in commodities in which a small state has a comparative advantage increases gains from trade but at the same time the small state becomes more vulnerable to external shocks such as a drop in export prices. Trade in manufactures and to some extent specialised services is generally of an intra-industry type but it is clear that conditions in small states do not allow the development of intra-industry trade and specialisation. This situation is not disadvantageous for small states although it limits the possible gains from economic integration as will be shown in chapter 9. To diversify exports and combat diminishing returns that usually characterises production based on natural resources, small states can develop activities based on human capital, production suitable for small firms or services such as tourism. Small states must also try to obtain foreign investment and attract industries owned by foreign firms to obtain technology and know-how. Location of economic activities and the effects of unequal economic growth is the subject of the next chapter.

Notes

1   Total consumption is estimated to be about 120 thousand tonnes in the US. Iceland exported 21 thousand tonnes in 1994. Source: Icelandic Freezing Plant Corporation, direct communication, February 1995.
2   Frozen fish fillets 11.7%. Salted, dried, and smoked fish, 11.6%. Source: United Nations (1993a).
3   This historical overview is based on Salvatore (1990:20–25).
4   For data on Denmark and the EU see Pavel Salz (1993:92), for Iceland: Þjóðhagsstofnun (1993:32, 92).
5   See OECD (1991:59).
6   A new quota system was put to use in 1991, see OECD (1991:30–33).
7   The most important demersal species are cod, haddock, saithe and redfish.
8   Total imports in SITC 03 (Standard Industrial Trade Classification) which includes fish, molluscs etc. amounted to 8,500 tonnes in 1989 (Landshagir, 1991:114).
9   Early criticism of the concept of intra-industry trade is that the level of intra-trade depends on the level of aggregation in trade statistics used to measure intra-trade. Grubel and Lloyd (1975:50, 67) analysed the aggregation problem on Australian data and found that intra-trade was reduced from 42% at the one digit SITC level to 6% at the seven digit level. They concluded, however, that intra-industry trade in Australia could not be explained away by disaggregation.

10   The weighted Grubel-Lloyd index is calculated from the following formula: $\Sigma(1-(ABS(X_i - M_i)/(X_i + M_i)))*(X_i + M_i)/(\Sigma X + \Sigma M)$ where X=exports, M=imports at the two digit SITC level where both imports and exports were recorded.

11   Approximately 1.5% of GDP originated in computer services and software production in 1991, this ratio seems to have doubled in 1993 (Þjóðhagsstofnun, direct communication, August, 1994).

12   Staffan B. Linder (1961) argued that nations tended to export, to nations with similar tastes and income levels, those manufacturing products for which a sophisticated domestic market existed. He was mainly concerned with manufacturing products but his demand oriented theory seems to be equally applicable to modern information industries.

8   Location and diffusion

In the previous chapter it was shown that the basis of trade in small states can be explained in terms of the Heckscher-Ohlin theory. Small states gain from international inter-trade by specialising in the production of commodities in which they have a comparative advantage or a relative factor abundance. All small states possess theoretically a comparative advantage in some products but this may not be sufficient to ensure an economic growth based on foreign trade if negative centre-peripheral forces lead to worsening terms of trade and draw skilled labour and capital from the small state.

Centre-peripheral forces affect all types of states or regions on the independence/dependence continuum. Small independent states have several means of combating polarisation effects that are not available in regions within larger states. The effect of these forces depends therefore on where the region or state is located on the independence/dependence continuum.

There are two main lines of thinking about the effect of economic or market forces in geographical space. One is based on classical economics and asserts that market forces will eventually reduce differences between states and regions, the other stresses the cumulative effect of growth in certain centres or growth poles, leading to increased disparity between regions or between centre and periphery.

The first question for small states, especially islands, viewed as peripheral regions, is whether strong 'backwash' or 'polarisation' forces are likely to draw people and capital from these places. This would imply a suboptimal system capacity and a reduced viability of small states. The answer given here is that small states can attach themselves successfully to the global economic system without fearing strong polarisation effects.

The second question is whether small island states can become an attractive location for industries and services. If conditions for the generation of growth poles or key industries are lacking in small states their international competitiveness would diminish. The answer is that a technical progress such as the lowering of transport and information costs have reduced the disadvantages of remote location and a small domestic market. Therefore it is possible for many industries to locate successfully in small states.

Centre-peripheral relations and unequal growth

Location of industry is governed by the local distribution of natural resources and other productive factors and the transportability of goods. Location is also a product of economic development (Ohlin, 1967:135). 'If a country has for one reason or another obtained certain manufacturing industries, technical labour will be educated and trained and other industries will be started because of its supply. Fixed capital takes on a technical form that affects location' (Ohlin, 1967:181–182). This statement hints at the cumulative nature of economic development. Factor mobility and commodity trade have a tendency to have an equalising effect although there is little tendency for domestic or international price differences to disappear completely (Ohlin, 1967:195).

If free trade equalised returns to factors of production, trade based on comparative advantage, with or without some factor mobility, would lead to similar rates of growth in small states or peripheral regions as in larger countries. The factor equalising theorem, which is deduced from the basic Heckscher-Ohlin model and the general equilibrium idea of classical economics, has been criticised as giving an incorrect description of the economic forces that govern economic growth and the location of industries.

Gunnar Myrdal (1957) puts forward the principle of cumulative causation. According to this principle the free play of the market forces normally tends to increase rather than decrease the inequalities between regions. Almost all economic activities that tend to give more than average returns in a developing economy will cluster in certain locations and regions, if left to free market forces, leaving the rest of the economy in the backwater. These clusters were formed at places that offered naturally good conditions in the beginning, but

> [...] within broad limits the power of attraction today of a centre has its origin mainly in the historical accident that something was once started there and not in the number of other places where it could equally well have been started, and that start met with success.

123

Thereafter the ever increasing internal and external economies interpreted in the widest sense of the word, to include for instance a working population trained in various crafts, easy communications, the feeling of growth and the spirit of new enterprise, fortified and sustained their continuos growth at the expense of other localities and regions where instead relative stagnation or regression become the pattern. (Myrdal, 1957:11)[1]

Expansion in one locality can therefore have 'backwash' effects in other localities. This means that the benefits to more developed regions will be at the expense of development in other regions. Rapidly growing regions will attract useful labour from other regions. Capital movements will increase inequality as increased demand will spur investment that will again increase income and demand and by that cause a second round of investment and so on. Savings will increase with higher income but will tend to lag behind investment because of the intensive use of capital. In other slow growing regions demand for capital will be relatively weak despite the low supply of savings from low incomes which tend to fall. Trade will also favour the more progressive regions that obtain decisive advantages in competition largely because of increasing returns. They also enjoy favourable terms of trade.

Myrdal (1957) also recognises certain 'spread' effects that counter the 'backwash' effects, for example regions around nodal centres of expansion should gain from the increasing outlets of agricultural products and be stimulated to technical advance. Regions that produce raw material for the expanding industrial centres will gain and their consumer goods industries will grow if enough workers are employed in the centres. Where the 'backwash' effects prevail the result will be a vicious downward growth spiral for the region and no equilibrium growth.

Nicholas Kaldor (1975) criticises the idea of general equilibrium in economics for neglecting the importance of increasing returns in industries. One consequence of increasing returns in industry is that industrial development tends to get polarised in certain 'growth points' or 'success areas' which become areas of vast immigration from the surrounding or more distant areas, unless this is prevented by political obstacles. This is one of the major reasons for the growing division between rich and poor areas of the world.

Kaldor (1970) argues that divergent regional growth cannot be explained by resource endowment except the part that consisted of 'land based' activities such as agriculture and mining. Only a small part of the interregional specialisation and the division of labour can be accounted for by such factors. The differences in real income between rich and poor regions or nations in the world are largely explained by the unequal incidence of development in industrial activities but not by 'natural' factors. Advanced high income areas have invariably developed industries that require capital, both machinery and human skills. Capital accumulation can, however, be a result of economic development as well as a cause of development. The growth of demand provides both the inducement to invest capital and the means to invest it as accumulation is largely financed out of business profits. Therefore industries will most often be located in regions that cannot be said to be well endowed with capital resources for reasons other than the industrial development itself.

Kaldor further argues that to explain why some regions become highly industrialised and others do not, the principle of circular and cumulative causation must be considered. This principle is equivalent to the existence of increasing returns to scale, in the broadest sense, in processing activities. The principle of comparative advantage and classical adjustment mechanisms do not work if trade is opened up between industrial and rural regions. The region with the more developed industry favoured by increasing returns will gain at the expense of the other region, and rich regions might obtain monopoly positions in industries that would eventually lead to the elimination of industrial centres in other regions. Lower cost industries tend to cluster in a number of successful regions holding each other on balance through increasing specialisation between them.

Empirical research related to disequilibrium theories

Recent empirical research indicates that regional growth in some industrial countries is converging rather than diverging. Robert J. Barro and Xawier Sala-I-Martin (1991:107), using regression analysis, have examined the growth and dispersion of personal income in the states of the US from 1880 to 1988 and gross state product from 1963 to 1986. Similar procedure was applied to 73 Western European regions from 1950–1985. The result was that for sectors and for state aggregates, per capita income and product in poor states tend to grow faster than in rich states. The rate of convergence is typically around 2% per annum. Other studies indicate that the growth of social services and producer services which have been the largest creators of new jobs in developed economies favour the most advanced regions and urban areas.[2] There is also a divergence in growth rates between individual countries and between major world regions in the period 1980–1989. East Asia, for example, grew 8.4% per annum and the industrial countries 3.0% while Sub-Saharan Africa grew only 1.0% and Latin America and the Caribbean 1.6% (The World Bank, 1990b:8). The question of divergent or convergent growth rates between regions, large or small, is not settled.

As far as the theory of cumulative causation depends on worsening terms of trade for primary producers its conclusions have not been borne out in reality. Kindleberger and Lindert (1978:75) state that except for a great fall in the terms of trade during the depression 1930, many primary products have faced favourable terms of trade for the last decades. 'The main problem has not been one of immiserizing growth in primary production. It is more likely to have been one of insufficient growth in primary production'. Table 8.1 shows that export prices of primary products are more unstable than prices of manufactures and prices in some categories have declined compared to manufactures in the period 1980–1989.

Table 8.1
Trends in export prices of primary products, 1970–1989, Index (1980=100)

Category	1970	1980	1989
Food	38	100	89
Beverages	32	100	61
Agricultural raw materials	30	100	113
Minerals and metals	45	100	119
Crude petroleum	5	100	59
Manufactures	34	100	120

Source:　　GATT (1990).

Table 8.2
Terms of trade in selected small states, 1968–1989, Index (1980=100)[3]

Country	1968	1980	1989
Barbados	53.9	100.0	72.6
Cyprus	198.4	100.0	109.2
Fiji	57.5	100.0	57.0
Gabon	32.5	100.0	53.8
Gambia	133.9	100.0	92.1
Guyana	123.0	100.0	78.2
Iceland	137.1	100.0	76.8
Malta	82.4	100.0	100.8
Mauritius	101.6	100.0	117.1
Seychelles	155.7	100.0	71.2
Trinidad and Tobago	72.9	100.0	55.4
Developing countries 1970–1989	50.0	100.0	84.0
Developed countries 1970–1989	121.0	100.0	112.0

Sources:　　The World Bank (1990a) and United Nations (1991).

Table 8.2 shows the terms of trade for selected small states and the developing countries in 1968, 1980 and 1989. Many developing states and small states mainly export primary or semiprocessed products. Several small states seem to have worsening or greatly fluctuating terms of trade during this period. The terms of trade have deteriorated for the developing countries in general since 1980.[4]

In figure 8.1 changes on the previous year in the terms of trade and GDP in Iceland are shown.

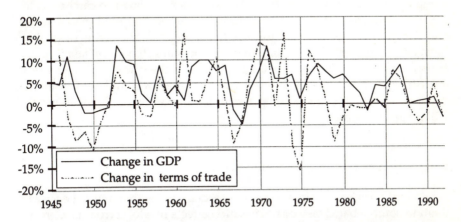

Figure 8.1   **Terms of trade and GDP in Iceland, 1945–1990, percentage change on the previous year**

Source: Þjóðhagsstofnun (1993).

In Iceland periods of slow economic growth are clearly associated with a downswing in the terms of trade (and *vice versa*). The terms of trade in Iceland that were favourable in the immediate post-war years began to deteriorate from 1948. There were three main reasons for this according to Ólafur Björnsson (1967). First, as the productive capacity of other fish-producing countries had been restored after the war the greater supply of fish products led to lower prices. Second, the development of the world market that followed the outbreak of the Korean War was unfavourable to Iceland as prices of imported goods rose considerably, without a corresponding rise in Icelandic export prices. Third, a dispute between Iceland and Great Britain about the fisheries limits off the Icelandic coast led to a close of the market for iced fish in Great Britain which caused a great deal of trouble for the Icelandic cod fish export.

The terms of trade improved considerably in 1953 but deteriorated again in 1957 for some years. In 1960 the terms of trade improved considerably and exports increased following good catches of herring until 1967. The terms of trade improved sharply in the beginning of the seventies as prices of primary products, especially food products, increased on the world market. A sharp deterioration in the terms of trade in 1975 and 1979 was connected with changes in the price of oil but economic growth was less affected than in the previous years when the terms of trade deteriorated. The present slump in the economy is connected with slightly worsening terms of trade.

It is clear that the terms of trade have been unstable in Iceland for a long time. It is, however, not possible to state that the terms of trade have been going against Iceland all the time. The swings are related to catches in Icelandic waters and supply of fish in foreign markets and food prices in general. Gross National Product follows the changes in the terms of trade in most cases although the economy has been less affected in the last two decades than previously. This seems to be due to the growth of new export industries such as tourism. If rapid population growth and environmental decay continues it is likely that prices of food and raw materials in general will go up in the future.

Polarisation effects and independence

Sovereignty brings certain advantages to small states that make it easier for them to combat backwash or polarisation effects resulting from fast growth of industrial areas in large states. The difference between independent states and regions within large states becomes especially important when advantages and disadvantages of regional integration are considered. A small state is less sensitive to international transmission of centripetal forces than regions within larger states as independence implies certain natural and man-made boundaries.[5]

> The boundary of a nation represents a point of discontinuity; it represents a change in the degree of mobility of almost all the factors of production [...]. To some extent these discontinuities are the result of real differences that follow national boundaries: differences of language, of education and skill, of a sense of community of outlook and interest. Such real differences are not capable of being wholly removed by the integration of the separate nations of today into larger units. But in great part the discontinuities are artificial. They derive from the existence of tariffs and other trade restrictions, of limits to the convertibility of currencies and the transfer of credits, of limits to the movement of labour or of other persons imposed either by governments or trade unions. (Robinson, 1960:xiv)

E. A. G. Robinson (1960:xiv) argues that the nation has obtained increased importance because it is the unit of government action and economic authority. Discontinuity at the boundary tends to become more pronounced as modern governments use the budgetary system as an instrument of economic policy within the national boundaries. Modern governments maintain full employment with the aid of a central bank, promote economic development and keep the balance of payments under control. Within the national boundaries the individual has rights to the benefits of the welfare state through social services and economic policy.

Albert O. Hirschman (1970) argues that both trickling-down effects (spread effects) and polarisation (backwash) effects are stronger in interregional than international relations. First, the mobility of factors of production is less internationally than interregionally. Second, nations compete in international markets on the basis of comparative advantage, regions within a country on the basis of absolute advantage. More efficient production in some region would eliminate similar production in another region but this does not happen between countries. Third, regions cannot protect their industries except through exemption from minor local taxes. An independent country can use tariffs to protect its industries. Forth, economic sovereignty in currency issue and exchange rate determination is an advantage for the independent state.

Hirschman adds that trickling-down effects are also stronger within states. Advances of one region will lead to purchases and investments in other regions as all complementaries that exist within a country will be readily exploited and regional specialisation emerges. Besides, regional policies are likely to be introduced if some regions are lagging behind. In an independent state adjustment is hindered by the 'friction of space', protectionism movements and reaction to balance of payments difficulties.

Geographical isolation hinders movement of factors of production in both directions in small island states. High transport costs strengthen the boundaries of small island states and increase the differences between a state and a region.

Polarisation effects in Iceland

In small states the most serious form of the backwash or polarisation effects occurs when workers, often highly skilled young people, choose not to stay in the region where they are born and raised but migrate to higher paying growth centres. Such 'brain drain' can happen in small states where the small home market cannot provide a sufficient range of specialised jobs.[6] Severe capital outflow is probably not a problem especially if the interest rate can be adjusted to foreign interest rates. In Iceland capital movements were restricted until Iceland became a part of the European Economic Area (EEA) in 1993. While interest rates are competitive in Iceland capital is unlikely to flow out of the country in large quantities. Competitive interest rates can, however, only be maintained in the long run if industries are competitive and profitable investment opportunities can be found. This in turn depends to a large extent on the availability of qualified labour.

The fishing sector has been the main growth sector in Iceland but there are signs of decreasing returns to investment and the fishing resource seems to have been fully utilised. It is therefore to be expected that unless new industries are built up, labour and to a lesser extent capital will move out of the country to find better opportunities in some faster growing regions. Figure 8.2 shows economic growth and migration to and from Iceland between 1961 and 1991. A clear trend does not emerge but labour from Iceland has emigrated to other countries in depressed times especially to Sweden when demand for labour in that country has been high.

Although independence clearly brings to the small state several possibilities to combat the backwash effects of fast growing regions in larger states, possible spread effects are at the same time reduced, especially if the state is geographically isolated (the area factor). Polarisation effects leading to the migration of labour are clearly visible in Iceland during times of economic depression but no lasting backwash effects can be demonstrated. Factors that directly affect foreign investment or determine whether industrial enterprises and service firms choose to locate in small states are probably more important than general polarisation effects.

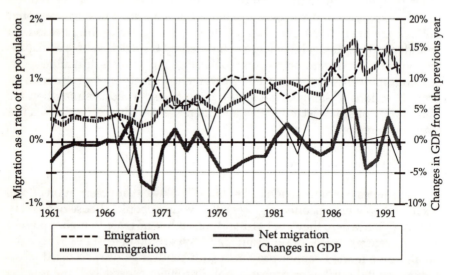

Figure 8.2 **Economic growth and migration to and from Iceland, 1961–1991**

Source: Hagstofa Íslands (1993).

Location of industries and growth poles

A direct attempt to connect economic growth to geographical space via the principle of cumulative causation is found in the idea of growth poles or growth centres. François Perroux (1970) observed that growth does not appear everywhere at the same time. The importance of innovation, the appearance and disappearance of industries and their different growth rates leads to the idea of a motor (or propulsive) industry. A motor industry is characterised by faster growth rates than the average growth rate for other industries and it takes the form of modern large scale industries sooner than others do. This means that individual factors of production are separated of from each other, capital is concentrated under one single power and labour is mechanised (Perroux, 1970:95). By applying the concept of external economies, industries (group of firms) are found not to be connected through the price system only, their profits are also induced by the inputs and outputs of other firms and their level of technique and results of research in production. New industries also help to create a 'climate' conductive to growth and progress.

Perroux then introduces the notion of key industry, a motor industry which by increasing its output induces growth in other industries inducing in the whole economic system an increase in total output much more than the increase in its own output. This is achieved by the motor industry by operating at increasingly lower points on its cost curve. Key industry and affected industries or clusters of industries often take on an oligopolistic form. By taking advantage of territorial agglomeration the non-competitiveness of the cluster is increased. These advantages are for example progressive consumption patterns of urban dwellers, location incomes added to business profits and overhead capital. Entrepreneurs and skilled workers have also influence.

Perroux's concept of growth poles leads to two problems. The first problem is how to identify a motor or key industry in an economic space, but this problem is not of particular interest for small states. The second problem is how a growth pole in economic space influences events in geographical space or leads to growth in certain places or regions.

Identification of growth poles

Growth poles are best regarded as sectors of an economy represented by an input-output matrix in which growth effects can be transmitted across the rows and columns. A firm or industry characterised by a high degree of dominance, great size and high interaction with many other firms is said to be propulsive (Darwent, 1969:6).

Hirschman (1970) considered two inducement mechanisms within the directly productive sector of the economy, forward and backward linkage effects. Backward linkage effects result as every non-primary economic activity will induce attempts to supply through domestic production the inputs needed in that activity. Forward linkage effects result as every activity that does not in its nature cater exclusively to final demands will induce attempts to utilise its outputs as inputs in some new activities. The degree of interdependence shown by any one industry can be measured by computing the proportion of its total output that does not go to final demand but rather to other industries and the proportion of its total output that represents purchases from other industries. From such computations it can be seen that for example the iron and steel industry has high backward and forward linkages but for services such linkages are low.

An industry with strong forward and/or backward linkages or a 'master industry' seems to correspond to Perroux's motor industry in many ways. Input-output analysis is, however, only able to determine a key industry *ex ante*. The impact of innovation can not be determined or predicted by looking at input-output tables. One reason is that although it is perhaps possible to estimate a final demand for the use of a product in its intended form and for its intended use, new products have been put into use never intended or suspected by its originators. In the case of service industries and industries based only on human capital or technological innovations this approach is also insufficient. The computer industry was identified early as a motor industry but the importance of the software production sector would not have weighted much in input-output analysis to start with. It seems sufficient to identify a key or motor industry simply as an industry that grows faster than the average growth rate of the economy and is not subject to diminishing returns in production.

Are motor industries likely to be found in small island states? The leading export industry in a small state such as the fishing industry in Iceland is likely to grow faster than other industries until the fishing stocks or other natural resources are fully utilised. The main export industry creates a progressive climate as it is in close contact with foreign markets. Strong forward or backward domestic linkages are, however, not likely to develop in small specialised economies. Export industries based on natural resources are subject to diminishing returns although this characteristic is not important if large oil reserves or similar resources are discovered in a small state. Fast growing services such as tourism are less supply constrained but increasing returns does not seem to be possible in tourism. There is also a limited possibility of growth of external economies when the economy is small and open. The fishing industry in Iceland has, however, helped to create new export products such as computer software and electronic equipment that is used in fishing and fish processing. These new industries are promising but the international competition is intense and their size is of little significance for the economy as a whole. It is difficult to see how a manufacturing motor or key industry with all the required characteristics can develop in a small island state but knowledge based service industries could develop in small states and become leading sectors.

Spatial effects of growth poles

The growth pole concept does not explain *a priori* the location of a propulsive industry in geographic space as D. A. Darwent (1969:8) points out.

The case of the multinational firm illustrates the problem of linking growth in economic space to geographical space. Their innovation power is strong due to high R&D spending. They have made use of sophisticated production methods that enable them to use unskilled or semiskilled labour with effect wherever it is cheapest. Their production and markets are widely dispersed in geographical space and they are also of a great size. Most multinational firms would therefore be regarded as motor industries in economic space but not necessarily key industries in any special region.[7]

Economic space can be conceived as a set of variables defined by economic concepts (price, elasticity) but geographical space as a set of variables that can be interpreted in locational context or corresponds to points on a map (distance, location). Connecting these two spaces requires that important variables from each space can go into relations with each other. In this way conclusions about economic processes in locational settings could possible be made.

The first variable to be considered is the distance variable. This variable is usually connected to the concept of transportation or transfer costs. Firms and industries tend to locate in such a way that transportation costs are minimised. One major criticism against using transport costs to determine the location of industries is the fact that transportation costs have fallen and are now probably well below 10% for most manufacturing industries and even less or non existent for service industries.[8] Information processing costs have also fallen rapidly. One estimation is that it has fallen by 65% from 1975–1985 (GATT, 1990:38). This makes it possible to locate activities dependent on information processing almost anywhere. Some of the most important potential growth industries today are therefore not affected by transport costs or distance factors.

Transportation costs are, however, substantial for many resource-based industries. That is the case in primary metal industries such as iron and steel industries. Iron and steel industries also have high backward and forward linkage effects. The ability of such industries to lower their costs could mean an important impetus for the whole economy. Such 'heavy' industries do not always have much regional impact. The aluminium industry has been rapidly growing in the last decades and is clearly a growth pole in economic space. Aluminium smelters require large amounts of electricity and they tend to locate close to sources of cheap electricity which is available in some developing countries as well as some developed countries such as Iceland and Norway. Further processing of aluminium takes place close to the market in industrialised regions but not necessarily where the smelters are located. The existence of an aluminium smelter in Iceland, for example, has not led to much further processing or localisation of processing factories. Iceland exports almost all of its primary aluminium production.

The second variable that could serve as a link between growth poles in economic space and geographical space is time. This variable is closely linked in many instances to the distance variable. For many (perishable) products the time it takes to transport the products to the market is important. In other instances the cost of waiting for spare parts' shipments, for example, matters more than actual transport costs. Also the speed of communication and information flow is of prime importance for industries and their management. Time saving is important for many industries and could be decisive in determining their growth potential. This is especially true of service industries. Technical innovations in communications and cheaper transport have reduced both time and distances for many practical purposes and because of that, time has lost its connection to distance to a certain degree.

The third variable is density of the population. This variable can be used to associate the concept of a growth pole with densely populated places or cities. Growth of service industries and cities is clearly related and a dense settlement increases the probability of entrepreneurs and innovations to appear. Although innovations can originate in any place the development of a marketable product from an innovation often requires extensive testing and stepwise improvements which depend on that many users have close contact with the product developer. The city provides the right environment for the development of products. Most importantly the city provides a supply of labour and services which are important to fast growing industries. But fast growing industries especially if they are based on natural resources are not necessarily located in big cities. Examples are the oil and metal industries.

Association of growth poles in economic space with a high density of population is reflected in the use of this concept in regional and industrial policies. Growth area policy as used in depressed regions involves selection of a few small areas that are considered capable of rapid growth. Effort is directed at building up and developing those selected areas. Development effort is therefore not spread evenly over a whole region nor necessarily directed into the most depressed areas. This leads to an increase in

134

the average size of settlements. There are, however, several diseconomies connected with cities and densely populated areas such as pollution which limit the usefulness of such policies. It is quite possible that many small sparsely populated regions and states will be able to offer a pollution free environment in which firms in finance, banking and other such services may prefer to locate in.

No strong locational forces seem to exist that make the location of many types of industries or services impossible in small states. The basic requirement to attract foreign investment or firms to locate in small island states is the ability to offer qualified and educated labour force. Investment in human capital is required to make a small state a growth pole.

The competition for foreign investment

Growth based on natural resources is subject to diminishing returns and cannot be the sole basis of growth and income forever. Small states must therefore try to diversify their economies by establishing manufacturing industries and service industries. Development of modern industries and services requires investment, marketing, skilled personnel and management. Capital and skilled labour are mobile factors for which the small states must compete internationally. If backwash effects predominate the small state will not be able to obtain foreign investment or foreign capital in sufficient quantities. Locational advantages or disadvantages are part of the cost or benefit of real independence of small states.

The capital for investment comes from three sources mainly; domestic savings, direct foreign investment and foreign loans. Developing countries may also obtain direct financial assistance from international aid programmes. Growth based on domestic savings alone would be very slow in small states as small states are suboptimal as a source of savings especially for large scale projects that are indivisible such as hydro-electric power plants. It would require very high interest rates and many years before capital for such projects could be raised domestically even in a high income state such as Iceland. Foreign capital has in fact been a *primus motor* in the economic progress of Iceland from the beginning of this century when the Bank of Iceland was established (1904) with a foreign share capital equivalent to 3 million Icelandic krónur which amounted to approximately 10% of GNP at the time. This capital was used to buy trawlers and modernise the fishing industry in Iceland which enabled rapid economic growth in the following years (Björnsson, 1994). Since then foreign capital has mostly been obtained in the form of foreign loans. Borrowing abroad is a limited source of finance and if it is not wisely used it can be disastrous. Foreign long term debt increased from 41.2% in 1988 to 62.8% of GDP in 1993 and 27.6% of export income 1993 was required to service this foreign debt

135

(Seðlabanki Íslands, 1994:19). Direct foreign investment is preferable as a source of economic growth if the debt/service ratio is getting too high.

G. K. Helleiner (1982:167–170) finds that there is a 'small country effect' in official development assistance and also a bias in favour of small countries (population less than five million) in non-aid capital flows, measured as average per capita capital flows and stocks in developing countries. The reasons are not clear but whatever the reason it seems that backwash effects are not found in the case of capital flows to small states.

A kind of growth pole strategy has been used by small states to obtain foreign investment. This is the widespread and increasing use of (Free) Export Processing Zones (EPZ) to attract industry. Such EPZs have been established in several small states such as Barbados, Belize, Bermuda and Malta. Other states such as the Cayman Islands have established tax heavens. Industry is attracted to EPZs by the use of tax concessions, easy credit, cheap labour and exemption from import duties etc. The benefits of EPZs for industrial development is highly uncertain. Dommen and Hein (1985:160–162) argue that EPZs have contributed positively to export performance of many islands but domestic value added and net export earnings tend to be limited to wage bills and some local services, seldom exceeding 25% of gross export figures. Linkages with other sectors in small island economies are extremely limited. These industries, often controlled by multinationals, are also notoriously 'foot-loose' and relocate easily. It is debatable whether the capital invested directly in infrastructure and indirectly in various concessions offered in EPZs is not better utilised in direct investment in local industries.

Although establishment of an EPZ has been contemplated in Iceland general measures to attract foreign investment have not been implemented.[9] Only in connection with energy intensive industries have direct attempts been made to attract foreign investors. Foreign investment is not allowed in the fishing sector, because of the fear of losing control over marine resources. This sector is probably of most interest to foreign investors. Instead of heavy borrowing abroad, Iceland must try to obtain direct foreign investment, otherwise the debt burden could reduce real independence drastically in the nearest future. For the purpose of obtaining foreign capital there is no reason why foreign investment should not be allowed up to a certain limit in the fishing industry.

Conclusion

Centre-peripheral forces affect small states, viewed as regions in the global economy, in similar ways as peripheral regions are affected by fast growing regions in larger states. The independent small state has more options to combat such forces than regions. The available data on the terms of trade and capital flows indicate that small states are not seriously affected by negative polarisation effects. In Iceland there are also no signs of strong polarisation effects from other countries or regions except for increased emigration in times of economic depression. Continuing growth and development requires effort to diversify the economy by creating jobs in services and small scale manufacturing industries. The small state must compete internationally for foreign investment to avoid excessive build up of foreign debt. Advances in communication and the lowering of transportation costs have made it possible for the small island state to become an attractive location for many types of services and manufacturing industries. The only prerequisite is investment in human capital.

A small state becomes more vulnerable to backwash effects of fast growing regions if it enters an economic union in which it has no political power. Regional integration is the subject of the next chapter.

Notes

1   Krugman (1991:61) gives examples of industries that have started because of 'an accident' in a particular location and 'thereafter cumulative processes took over'.

2   See UNCTAD (1988:147–148).

3   The ratio of export prices, FOB and import prices, CIF.

4   Salvatore (1990: 320–321) discusses studies on the movements of both commodity and income terms of trade. Although the commodity terms of trade may have declined, the income terms of trade of developing countries have increased due to an increase in the volume of trade.

5   Nation and state are here assumed to coincide.

6   Reliable data on emigration in small states seems not to be available but there are indications that migration rates are higher in island states than continental states. See Cleland and Singh (1980:971).

7   Lall and Ghosh (1982:153–156) discuss the role of multinational companies in technology transfer to small states.

8   See endnote 5, chapter 5.

9   See Byggðastofnun (1987).

9   Regional integration

Integration into the global economic system helps a small state to overcome the disadvantages resulting from a suboptimal domestic market and increases system capacity as well. Regional integration is a more direct way to escape disadvantages resulting from a suboptimal size of the market and in addition it extends the boundaries of the political system. As an alternative to independence regional integration reduces the cost of real independence. The question of regional integration is especially important if a small state is within the sphere of influence of a geographically close union which happens if, for example, the main trading partners are inside the union. The small state would then have no part in decision making within the boundary of this extended political system, unless it obtained membership. The power of the small state and, indirectly, citizen effectiveness, is increased by integration in this case as the small state can now participate in decision making for the whole region or union.

Is there a strong tendency for a free trade area to evolve into a customs union? If there is a strong tendency to either increase or decrease the level of a particular form of regional integration the integration is unstable. By the final stage of economic integration (total economic union) a new state, probably a federal or a confederate one, has been formed. This leads to the related question of the stability of very large states. Do they have a tendency to break up into smaller units? What is the optimal size of a state from the point of view of stability? A full answer to these difficult questions will not be provided here but it seems that the lower forms of integration (free trade areas and customs union) are not stable when tried in developing countries.

An important question is whether small states obtain larger economic gains from regional integration than from free trade on the world market.

The answer given here is negative. The conditions for intra-industry trade are limited in small states especially small island states that are confined to inter-industry trade and this situation is unlikely to change in the future. The dynamic benefits of economic integration are mostly due to an increase in intra-industry trade. Small states will, therefore, only be able to benefit from static trade creation effects when entering a customs union and such effects are likely to be small or of limited duration.

Are there other benefits than economic from regional integration that could increase the system capacity of small states? The answer is that a small state might increase political power by participation in a customs union or higher types of integration only if each participating state has equal rights, independent of size. Otherwise the small state might practically degenerate into a peripheral region in the union.

The final question is how Iceland could benefit by joining the European Union. It turns out that Iceland is unlikely to gain much economically in addition to the gains already obtained, especially by tariff reductions, through the European Economic Area treaty. Iceland can increase its political power by joining the EU, especially through cooperation with the other Nordic countries that are members of the EU (except Norway).

Independence and regional integration

Development from one type of integration into another corresponds to a movement along the independence/dependence continuum; from independence to integration or *vice versa* if integration becomes less intense. International treaties require the participating states to surrender their independence or national sovereignty to a certain extent but since independence in one area is then traded for other benefits which strengthen the economy or the social system in other areas real independence may be increased. A small state that wishes to participate in a free trade area or a customs union may be confronted with the choice to surrender independence partly, both in a real and formal sense, or else withdraw at some stage from the integration process. If there is a strong tendency for some forms of economic integration to call for more and more integration until total economic union has been achieved or a new state has emerged the small state would become a region within that new state.

The development from a customs' union to an economic union was deliberately planned from the beginning in the case of the EU and more integration is planned (towards a Federal Europe?). The basic structure of EFTA remained mostly unchanged for thirty years but then it 'developed' into the European Economic Area and the member states except Iceland, Norway and Switzerland have now entered the EU.[1] The experience in other parts of the world seem to indicate that free trade areas are rather unstable and tend to break up, especially when states of different size and economic development try to integrate.[2] The experience in Europe indicates that economic integration, if it progresses beyond the free trade area stage, tends to progress further towards an economic union or even a total union. A small state might therefore expect to loose formal and possible also real independence if it participates in a regional economic integration. In the case of Iceland joining the EU it would become a peripheral region in Europe. Cooperation between states and a high level of cultural and social integration is possible, however, without progressively increased political and economic integration. The Nordic countries have a long history of close cooperation in various cultural and social affairs. They have established, for example, a common labour market legally. No economic union has evolved although the formation of such an union has been suggested several times. Small states should therefore not hesitate to seek cooperation with other larger countries in areas of cultural and social importance.

The theory of economic integration

Economic integration is possible in many ways. The most important types of integration are listed below very briefly:

A *free trade area* is an agreement among countries about elimination of all tariff and quantitative restrictions on mutual trade. Every country in this area retains its own tariff and other regulation of trade with third countries. The bases of these agreements are the rules of origin. These rules prevent trade deflection that is the import of goods from third countries into the area by country A (which has a relatively lower external tariff than country B) to re-export the goods to country B.

In a *customs union* participation countries not only remove tariff and quantitative restrictions on their internal trade, but also introduce a common external tariff on trade with third countries. The participation countries take part in international negotiations about trade and tariffs as a single unity.

In a *common market,* apart from a customs union, there exists free mobility of factors of production. Common regulations (restrictions) on the movement of factors with third countries are introduced.

An *economic union* among countries assumes not only a common market, but also the harmonisation of fiscal, monetary, industrial, regional, transport and other economic policies.

A *total economic union* among countries assumes a union with a single economic policy and a supranational government of this confederation with great economic authority. (Miroslav N. Jovanovic, 1992:9)

Static effects - trade creation and trade diversion

The static welfare effects resulting from the formation of a customs union (or a higher order union) can be analysed in terms of trade creation and trade diversion. Trade creation represents an improvement in resource allocation but trade diversion represents a worsening in this allocation.

The institution of tariffs affects resource allocation in two ways according to Bela Balassa(1973:21–29): (a) the production of some commodities will shift from lower-cost foreign producers to protected home producers operating with higher costs (discrimination against foreign sources of supply of the same commodity); (b) consumer demand will shift from foreign goods to domestic products in response to the change in relative prices consequent upon the tariff (discrimination against foreign goods that are different in kind from domestic goods). Whether the net effect for the union's establishment leads to freer trade or increases discrimination depends on the relative magnitudes of production effects, the consumption effects, the terms of trade effects, and administrative economies.

Trade creation represents a movement towards the free-trade position, since it entails a shift from high-cost to low-cost sources of supply, while trade diversion, a shift of purchases from lower-cost to higher-cost producers, acts in the opposite direction. Beneficial effects will predominate if trade creation outweighs trade diversion. But the outcome depends on unit costs. The lifting of quantitative restrictions will always have a positive production effect. World real income can also increase in the absence of any improvement in productive efficiency, provided that efficiency in exchange improves. It is also possible that a customs union will result in negative production effects but this loss in welfare will be more than outweighed by the gain in consumer satisfaction derived from the abolition of discrimination between domestic commodities and the products of the partner countries (Balassa, 1973:67).

A customs union is more likely to lead to trade creation and increased welfare under the following conditions *inter alia* (Salvatore, 1990:293–294):

1   The higher the preunion trade barriers among member countries.

2   The lower the tariffs with the outside world.

3   The wider the union or the greater the number of countries forming the union. In this case there is a greater probability that a low-cost producer is found within the union.

4   The more competitive rather than complementary the economies of member nations are. More competitive means that there is greater number of similar goods produced in the participating countries. Due to differences in productive efficiency, each country will expand its comparatively more efficient industries and contract the less efficient ones. A complementary union between for example an industrial nation and agricultural nation would be less likely to lead to increased efficiency.

5   The closer geographically the members (lowering of transportation costs).

There are also administrative savings resulting from the elimination of customs officers, border patrols, etc. Collective terms of trade are also likely to improve if a trade-diverting customs union reduces its demand for imports from the rest of the world and reduces its exports. For a trade creating customs union the opposite is the likely case. The bargaining power of a customs union can also be greater than in the case of each member separately.

In a study on trade creation and trade diversion in the European Common Market, covering the period from 1959 to 1970 approximately, it was found that benefits from economies of scale, the rationalisation of production and increased investment are more important than static gain or losses from trade creation and trade diversion. This also applies to benefits of non-member countries who might gain from effects of economic growth in the actual member countries. Trade creation was found to have significant effects and be more important than trade diversion. In addition it was found that trade creation is mainly the result of intra-industry specialisation in manufacturing based on economies of scale (Balassa, 1974:126–127).

Dynamic effects and benefits

The dynamic effects of a customs union are due to increased competition, economies of scale, stimulus to investment and better utilisation of economic resources. These effects are the result of the expansion of the market in one or the other way. Direct connection between productivity levels and the size of the market is not easy to measure. Measuring the size of the market itself is also a problem. The size of the GNP seems to be the most useful measure although the size of the population may be a better measure in some instances and GNP per head in other.

Bela Balassa (1973:108) advances the hypothesis that with given natural resources and capital, a higher level of manufacturing productivity can be attained in a wider market. From this it follows that economic integration leads to an improvement in the dynamic efficiency of the participating nations. Empirical studies relating to this hypothesis include the findings of P. J. Verdoorn who found that there was an interrelationship between the growth of output and the rate of increase in productivity over time in a number of industries in advanced nations, in the period 1870–1914 and 1914–1930. Balassa (1973:115–116) concludes:

> These findings indicate that the size of the market is an important variable in determining the level of productivity. This does not mean, however, that the impact of factors other than market size could not outweigh the advantages of a larger market. High levels of productivity in small countries, such as Belgium, Switzerland, or Sweden, require the introduction of further explanatory variables. A stronger sense of community, a higher standard of education, and a greater capacity to adjust are often cited to explain the high per capita incomes in these countries.

Balassa adds that sociological and psychological factors also play an important role in small and large countries and the presence of non-economic factors may affect the conclusion that a wider market will make possible the attainment of higher levels of manufacturing productivity and that a fusion of national markets will improve the growth prospects of the participating countries.

The gains from a greater market size are obtained through economies of scale in industries. Economies of scale of a technical origin have been studied in connection with the completion of the internal market of EU in 1992. The economies of scale in a given sector are appraised empirically using the concepts of minimum efficient technical scale (METS) and cost gradient that represents the increase in unit output costs where the firm is of less than the optimum size. The result of a recent study indicated a rise in unit costs for firms below METS (Emerson et al. 1988:128). Economies of scale were found to be relatively large in transport equipment, chemicals, machinery and instrument manufacture and paper and printing. These sectors are important in the economy of the EU as they account for 55% of industrial production in the Community. They are also sectors where demand is often growing strongly and their products have a high technological content. This applies to office machinery, electrical and electronic equipment, precision instruments, chemicals, pharmaceuticals and products of the aerospace industry. In food, drink and tobacco, textiles, clothing, leather goods and timber, economies of scale are less important. The sectors with small economies of scales are characterised by a relatively stagnant demand and a low technological content of their products (Emerson et al. 1988:128).

Insurance, banking and distribution services are also traded internationally. Trade in these services has been growing fast in the last decades. Despite its growing importance it seems that trade in services has not been investigated or analysed as much as trade in goods. It seems, however, that economies of scale play a major role in many service industries.[3]

Non-technical economies of scale relate to economies at the level of the firm in sales promotion, R&D, management and financing and the level of transport costs. In finance, a 4% difference in interest rates paid by small businesses and the largest firms has been found. Economies of scale were not found to exist in the innovation process but effects of experience and learning were important. Unit costs falling by 30% in sectors like electric components and micro computing have been reported (Emerson et al. 1988:138).

Market integration makes greater technical efficiency possible. The size of the market has a significant effect on the size of production units and the expansion of an industry's market through foreign trade is generally accompanied by a significant increase in the average size of production units. Reallocation of resources takes place between countries in favour of those enjoying a comparative advantage. Such restructuring looks possible only in sectors where the minimum technically efficient scale is large in relation to the domestic market (Emerson et al. 1988:140–144).

Various factors can contribute to the increase in competition in an integrated area. Balassa (1967:165) argues that even if a small country were able to support firms of optimum size in various branches of manufacturing, it would rarely be possible to have a number of efficient firms competing in a small market. The removal of tariffs increases the number of potential competitors and will thereby loosen monopolistic and oligopolistic market structures. In sectors characterised by national monopolies, oligopoly will become the dominant market structure, while in oligopolistic industries the size of the group will increase.

Besides the direct effects on prices increased competition may promote innovation and a technical progress. The completion of the European internal market should have a positive overall effect on innovation due to increased competition, more openness to international trade, increased growth potential and intensification of technological development by increased mobility of researches (Emerson et al. 1988:162).

Small states and regional economic integration

Intra-industry trade is negligible in the foreign trade of the majority of small states. Foreign trade of small states is now characterised by inter-industry trade and foreign merchandise trade consists mainly of agricultural products or raw materials such as oil. These products have the character of primary production. Producers of primary commodities are mostly confined to inter-industry trade as primary production takes place under decreasing returns to scale. A larger market is therefore not necessarily of interest for the primary producer unless higher prices can be obtained.

Tariffs, tariff equivalents or other trade restrictions are still high in many areas of trade in agricultural products and some other primary products. Trade in manufacturing products such as textiles is also subject to restrictions. The Uruguay round of GATT negotiations addressed these problems. Tariffs and restrictions on trade in products of agriculture will be lowered in the next years but tariffs will continue to be substantial in many areas.

Many small states specialise in agricultural production for export and trade diversion and trade creation effects of regional integration are therefore of importance to them. A small state may become a victim of second best policies resulting from the formation of a customs' union in the sense that trade diversion may severely hurt the export industries of a small state with no or little bargaining power outside or inside such a union. The EU common agricultural policy has without doubt led to a trade diversion that for example hurt countries like New Zealand when Britain joined the EU (then EEC). Trade diversion and trade creation are unlikely to be of much importance for most manufacturing industries. During the last decades tariffs on manufactures have been lowered under the GATT agreement and are now around 5% in industrial countries, but non-tariff measures to limit trade have replaced tariffs in some cases.[4]

There may be static gains from trade creation in agriculture and some primary production if a small state enters a customs union. As lower tariffs and new rules for trading in agriculture and primary products in GATT have been accepted, such static gains are not likely to be important much longer. The small state may as likely miss some opportunities in other markets because of its association with the union's partners.

This conclusion applies to the small states that are now dependent on primary production to a large extent. The conditions for increasing return activities based on product differentiation, large scale production and high R&D content does not seem to exist in a small state. In the case of small island states foreign trade will help to increase the size of the market and a high level of education and training may make a small state a desirable place for the location of high technology industries. The essential thing is

145

that the small domestic market makes it impossible for the small state to absorb imports in sufficient quantities to enter the reciprocal (merchant-producer) relationship required for intra-industry trade.[5] In short the small state will always be a marginal market from the point of view of large scale enterprises.

The case of Luxembourg shows how a favourably located small state can take advantage of trade and production of service industries. Not much is known about international trade in services but it seems that services have certain characteristics of both increasing and decreasing cost industries. The most obvious difference between services and manufacturing production is that services are not produced for stock and transport costs are negligible. They are human capital intensive and require good communications and close contact between the producer and consumer. Information technology has made the location of many such activities flexible and it is quite possible that many small states can develop export industries in services like banking and insurance (some small states have become tax havens and centres for off shore banking but such activities seem not to develop much domestic value added or local multiplier effects). Small states will probably benefit most from regional integration in service industries.

Iceland and European integration

Early in the 1960s a discussion started on the possible association of Iceland with the European Economic Community (now European Union) and EFTA. Iceland had to pay attention to developments in Europe, because of strong economic and cultural ties. During 1961 an Icelandic delegation had a detailed discussion with German representatives on the problems that Iceland faced after the creation of the EEC and possible actions that Iceland could take.[6] One estimation of the possible effects of the common tariff on Icelandic fish products, given that most of Western Europe would adopt the common tariff, was that the fishing industry would lose around 14% of net value of products (Jónas H. Haralz, 1962:11).

It was pointed out at the time that an association with the EEC would be a rather difficult process. This was firstly due to the extensive protection of Icelandic industry both in terms of high tariffs and quotas. Besides, trade with the Eastern European countries was based on import restrictions on commodities from other countries (this was a form of a barter trade). Therefore it was realised that a long period of adaptation was necessary. Secondly the principle of free movement of capital and labour between countries was considered impossible to accept in Iceland. As a very sparsely populated state with heavy dependence on fishing, Iceland would

146

require special permanent status in the European Economic Community.

> Therefore, the special provisions that Iceland requires with regard to labour- and capital-mobility and with regard to rights concerning the operations of business premises, must be without time limits. These provisions must ensure that not more of foreign capital and labour will be transported to Iceland than the Icelanders themselves regard safe for their nationhood and culture and for their control over the economy. They must also ensure that foreign fishermen will not be allowed to fish within Icelandic fishery limits and that possible participation of foreigners in fishing enterprises will not lead to depletion of fishing stocks in Icelandic waters. (Haralz, 1962:20)

This point of view which was taken up by the Icelandic government in the sixties towards the EEC and EFTA has not basically changed since.

The response of the UK and the Nordic countries to the EEC was considered to be of special importance. Interest in EFTA diminished when it became clear that the EFTA countries would negotiate individually with EEC, but it was at that time believed that most members of EFTA would be a part of the EEC. In 1963 it became clear that the UK would not be a part of the Community for the time being. Good fish catches, especially herring catches, which fuelled rapid economic growth made the economic effects of European Integration less felt in Iceland for some years. In 1967 and 1968 Iceland experienced a large drop in fish catches that coincided with a fall in export prices. Unemployment rose and the country experienced heavy emigration to Sweden, Australia and some other countries. This situation stimulated interest in finding ways to diversify the economy and making it less reliant on fish and fish processing. The EU customs union and EFTA free trade area were especially effective in terms of manufacturing trade. The development of manufacturing industry in Iceland was therefore considered to be difficult without some association with either or both of these unions. As EFTA was the main trading partner and its obligations were limited to trade, Iceland sought association with EFTA. Iceland joined EFTA in March 1970. Protective tariffs on Icelandic manufactures were abolished in the EFTA countries at the moment of ratification. Iceland's protective tariffs on EFTA industrial products were lowered by 30% immediately but remained unchanged for four years thereafter. From January 1974, protective tariffs were to be lowered annually by 10% until reaching zero rates in January 1980.[7]

After the association of Britain and Denmark with the EU, Iceland wanted to continue her important free trade relations with those two states as well as to obtain free trade with other member countries. A free trade agreement with the EU was reached in 1972 and came into effect in March

1973. The pattern of tariff cuts was to follow the same path as agreed in the case of EFTA. An important part of this agreement is Protocol 6 that allows the special treatment of fish exports from Iceland to EU. Most fish products were not classified as manufacturing products in the general free trade agreement between EFTA and the EU. Several important fish products were excluded from Protocol 6, notably fresh fillets, salted fish and herring. Icelandic producers were hard hit by tariffs on salted fish when Portugal and Spain joined the EU in 1986. It has been estimated that tariffs paid on Icelandic fish imports to the EU were 3.5% of total sales in 1990 but without the special agreements with the EU this sum could have been 6%-7% of exports (Hagfræðistofnun Háskóla Íslands, 1991b:25).

Trade with the EU increased greatly after 1981. In 1981 26% of marine products went to the EU but in 1989 around 60% of marine exports and 50% of industrial exports went to the EU (Þjóðhagsstofnun, 1991c:23). Further steps towards formal economic integration were, however, not taken until discussions on the European Economic Area (EEA) started in 1989. A treaty was ready for ratification in 1992 and was originally supposed to come into effect in January 1993 but when Switzerland rejected it in a referendum, changes had to be made and the treaty came into effect later that year. In this treaty tariffs on several articles left out of Protocol 6 are greatly reduced or abolished. Trade in fish is still not as free as trade in manufactures. Fishing and fish processing is subject to a special fisheries' policy in the EU in line with the agricultural policy. Various forms of subsidies and regulations strengthen the competitive position of the fish processors especially, and also the fishermen in the Union compared to the Icelandic producers. The industry most affected by the EU tariffs is probably canned fish and fresh fish products. In a recent study it is stated that the present rules on trade in marine products and fishing rights create substantial costs for the Icelandic economy.[8] While fisheries are not a part of the EU/EEA free trade arrangement, possible gains for the Icelandic economy by the agreement will be much less than the gain for those states where manufacturing and services are the main industries (Hagfræðistofnun Háskóla Íslands, 1991b:11). The subsequent negotiations with the EU led to a reduction of tariffs on most fish products. On the average tariffs are to be reduced by 70% in January 1997 (Utanríkisráðuneytið, 1992:10). Although the lowering of tariffs will increase prices, the gains are likely to be divided between producers and consumers.

One of the main ideas behind association with EFTA was the desire to diversify the Icelandic economy although free trade with fish is also of primary importance (totally free trade with fish was not achieved within EFTA until 1990). One area, which looked promising, was the development of energy intensive industries such as aluminium smelting. A contract was made in 1966 with Swiss Aluminium to build the aluminium smelter that started to operate in 1969.[9] In 1969 the share of industry in exports was around 10% (5% excluding aluminium). This ratio rose to around 24% in 1976 (7% excluding aluminium). Annual growth of industrial production was around 6.5% per annum (aluminium excluded) while GNP increased by around 4% during this period. Financial status of firms that aimed at exports was difficult in the first years after the association with EFTA, but improved greatly in 1977 compared to previous years and other industries. (Þjóðhagsstofnun, 1977:16). This favourable trend in industrial production did not continue. The share of manufacturing industry in merchandise exports was only 17.4% in 1991 (without aluminium exports 8.4%). In table 9.1 the changes in main exports from 1977 to 1991 are shown.

Table 9.1
Exports of selected manufactures in Iceland, 1977 and 1991,
millions of krónur, current prices[10]

Industry (Icelandic classification)	1977	1991
Nonalcoholic beverages	-	177.4
Alcoholic beverages	-	7.4
Seaweed meal	-	63.1
Fish feeds	-	456.9
Diatomite	830.2	420.3
Fish tubs, netrings, etc.	-	140.4
Products of tanned skins	157.8	-
Tanned or dressed skins	1092.5	832.0
Packing containers of paperboard	-	59.0
Wool tops and wool yarn	605.8	88.4
Woollen fabrics	-	20.4
Fishing lines, cables and nets, etc.	-	110.9
Knitted clothing, mainly of wool	2,406.0	564.7
Other garments	115.7	59.1
Woollen blankets	264.2	86.1
Rock wool	-	67.9
Aluminium pans	-	286.3
Electronic weighing machinery	-	258.9
Equipment for fishing	-	150.2
Fish processing machinery	-	187.4
Other manufacturing products	730.4	706.8
Total	6,202.6	4,744.0
Total exports of manufactures	22,342.6	15,924.2

Source: Hagstofa Íslands (1978) and (1992).

This table shows that a few new industries have started to export in 1991, mainly products related to fish and fish processing. The textile industry was important in 1977 but went into difficulties in the nineties. One of the reasons was that important markets in Eastern Europe collapsed. It seems that tariff free access to the EFTA and the EU market has not resulted in substantial growth of the Icelandic manufacturing industry. Also from the point of view of diversifying the economy not much progress has been made.

The EEA treaty provides free trade in manufactures, free trade in services and free mobility of capital and labour. By joining the EEA, Iceland has removed as far as is possible man-made restrictions on the size of its home market. The long distance from Europe can, however, not be changed . The CIF/FOB ratio can be used to measure approximately the cost of the economy due to isolation (or distance-protection of domestic producers). This ratio was 11.4% for Iceland in 1985–1988 as compared to 3.8% for Denmark, 2.5% for Norway, 2.6% for Sweden and 4.3% for Britain (Hagfræðistofnun Háskóla Íslands, 1991b:71). This ratio varies greatly between products and limited competition in transport may add to higher transport costs.

Over 70% of Iceland's merchandise trade (imports 71%, exports 73% in 1991) is conducted with countries belonging to EFTA and EU. Therefore static trade diversion or trade creation effects resulting from Iceland's association with the EEA are unlikely to be significant. However some fish exports now going to the US might be diverted to the EU if all tariffs on fish products were eliminated between Iceland and the EU. This might lead to trade creation because Icelandic fish production is considered to be more efficient than the EU fish production. Consumption effects would probably be most noticeable in agriculture. Icelandic agriculture has been isolated from external competition for a long time and prices of domestic agricultural products are usually higher in Iceland than in the EU. This isolation will be broken gradually when the recent GATT agreement on agriculture comes into effect.

Dynamic effects from increased competition and economies of scale in industry are difficult to access. The average size of Icelandic firms is very small in terms of the number of employees. Only 8% of industrial firms have more than 20 employees but in Northern Europe 75%-90% of industrial firms have more than 20 employees and around 45% of firms had over 100 employees (Hagfræðistofnun Háskóla Íslands, 1991b:66). Only two industries in Iceland are considered 'sensitive' to competition from the internal EU market, namely ship building and ship repairs (ISIC 381) and beverages (ISIC 213) having a combined share of 16% of gross factor income in industrial production in 1987. The most likely dynamic effects are connected with a more open capital market resulting in a more stable exchange rate and interest rates closer to the rates available in Europe.

There is also a greater likelihood of foreign enterprises willing to locate factories in Iceland or invest in Icelandic companies thereby creating larger and more competitive units.

Other benefits of regional integration

The inter-industry nature of Iceland's exports makes it unlikely that Iceland will gain much in direct economic terms by entering the EU. The GATT treaty and the EEA treaty seem sufficient to ensure adequate market access in the long run and other possibilities such as the North American Free Trade Agreement (NAFTA) could be as advantageous in terms of direct economic gains. The case for Iceland joining the EU seems to rest on political status and the amount of power that membership entails. Most of Iceland's trade goes to the EU. Culturally and socially Iceland is tied to the Nordic countries that have, except for Norway, entered the EU. It is clear that the European Union will be the forum of discussion and decision making for Europe in the future. Whether Iceland likes it or not it is within the sphere of influence of all decision making in the EU, especially since the EEA treaty came into force and the other Nordic countries (except Norway) became members of the EU. If Iceland wants to have some political influence in Europe it must join the EU.

The gains in real independence are dependent upon the future treatment of small states and the interpretation of the subsidiarity principle. Given that all independent states of the EU are treated equally without regard to size in terms of population, Iceland will obtain much more political power than the relative size would permit if it stayed out of the EU or were not a sovereign state. Citizen effectiveness would increase indirectly as the representatives of the people would have more power than before. System capacity might also increase as it would be easier to obtain capital for an economic development. The situation would be much less favourable if small states were treated more or less as regions in the EU. Centripetal forces could be more difficult to fight and citizen effectiveness would probably not increase.

The subsidiarity principle ensures that if the member states cannot attain the objectives of a proposed action the Community will assist. This principle can be interpreted radically as giving almost full autonomy to lower levels of government within the EU (Rodríguez, 1993). This interpretation, if accepted, gives a small state that is contemplating to become a member of the European Union some assurance that the present level of real independence will not be much reduced within the European Union.

Conclusion

A small state may try to increase its system capacity by regional integration. Economic gains from integration are probably less than in the case of large states. The main reason is that conditions for intra-industry trade are limited in small states. Therefore small states will only be able to benefit from static trade creation effects when entering a customs union and such effects are likely to be small or of limited duration. Political gains through regional integration are only possible if each independent state has equal rights regardless of size, otherwise a small member might practically degenerate into a peripheral region in the union. Regions have fewer means to combat polarisation effects than independent states as shown in the previous chapter. Iceland, in particular, is unlikely to gain much economically by entering the European Union in addition to the gains already obtained by tariff reductions through the European Economic Area treaty. Iceland might increase its political power by joining the European Union if the present status of small states in the Union is not changed.

Notes

1   Switzerland is not a member of the EEA.
2   See Mario I. Blejer (1988) for a discussion of the regional integration experience in Latin America.
3   Michael E. Porter (1990:239–252) discusses service industries and trade in services.
4   See UNCTAD (1988:231).
5   See Meyer (1979).
6   See Gylfi Þ. Gíslason (1979:186).
7   See Jón Sigurðsson (1970:19).
8   Tariff revenue in the last years on Icelandic fish products is estimated to be around 2,000 million Ikr at the 1991 exchange rates (Utanríkis-ráðuneytið, 1992:11).
9   See Garðar Ingvarsson (1987:164–177) for details on the development of energy intensive industries and the use of thermal and hydro power in Iceland.
10   Prices in 1977 are in old krónur that are 1/100 of new krónur introduced in 1980, but the price index has increased approximately 100 fold 1977–1991. Use of constant prices would therefore not make much difference.

10 Conclusion

There are no serious disadvantages resulting from small size of states in the global system. If such disadvantages exists they have been largely overcome by several small states including Iceland and Luxembourg. Other small states can obtain a similar standard of living as Iceland. The only prerequisite is education and investment in human capital. It is clear, however, that small island states have a special position in the world system. This is caused by geographical factors, trade characteristics and the lack of political power.

Geographical factors, especially in combination with a small population, influence the economic development and political strength of small states. The effective size of the state is enlarged if the Exclusive Economic Zone is included as a part of the state but isolated or scattered settlement structure over a large sea or land area make the state effectively smaller and increase problems of the political system. Technological advances in communication coupled with suitable regional policies help to reduce the negative contributions of the area factor.

Trade characteristics associated with small states include a high ratio of trade to GDP on the average, a high content of primary production or natural resources in their exports and a high geographical concentration in exports. These characteristics have been verified for small states with population less than one million in this study. To explain these characteristics there is no need to go further than to the Heckscher-Ohlin theory of the basis of trade. Further analysis confirms that the most important trade characteristic of small states is that they are confined to inter-industry trade and specialisation and the conditions for intra-industry trade based on increasing returns activities are lacking. This does not mean that small states are unable to develop and compete internationally in technically advanced industries or services. Technical progress leading to a lowering of transport and information costs has reduced the disadvantages of a remote location and a small domestic market. It is therefore possible for many industries to locate successfully in small states.

Centre-peripheral forces affect all types of regions on the independence/dependence continuum. Small independent states have several means to combat polarisation forces that are not available to regions within larger states. This includes monetary and fiscal policies. Small states can therefore attach themselves successfully to the global economic system without fearing strong polarisation effects. This underlines the value of independence.

Small states are clearly suboptimal in terms of military power and do not have the resources to maintain adequate military defences. In matters of defence they are therefore necessarily free riders and must try to guarantee their security by bilateral or multilateral alliances. Stability in the global system is of utmost importance for small states as regional conflicts can easily threaten the very existence of small states especially those that are landlocked or semi-landlocked. Active participation in the United Nations and other international organisations is also necessary to further the objective of global and regional stability.

Regional integration is a direct attempt to increase political and economic power of small states. It is not likely, however, that small states will obtain larger economic gains from regional integration than from free trade on the world market. The main reason is that the conditions for intra-industry trade are lacking in small states but dynamic benefits of economic integration are mostly due to the increase in intra-industry trade. Small states will only be able to benefit from static trade creation effects when entering a customs union and such effects are likely to be small or of limited duration. The case for regional integration of small states rests therefore on political advantages. But small states increase their political power only if each independent state has equal rights in a regional union, independent of size. Otherwise the small state might practically degenerate into a peripheral region in the union.

Elements of a theory of the optimum size of states have been developed in this study. Further research is needed on optimum criteria, constraints on optimality and real independence to complete such a theory. It is also necessary to investigate how centre-peripheral forces and differences in economic development affect the location of states and regions on the independence/dependence continuum. The difference between regions, independent small states and special status regions must be further clarified. Other areas of further research on small states could focus on the theory of intra-industry trade to clarify the status of inter-industry traders and the role of the size factor in economic development which is an important topic. A comparative study of Iceland and a less developed small state such as the Fiji Islands might clarify the problems of underdevelopment in small states. Other possibilities include comparative studies of an independent nation-state such as Iceland and peripheral (problem) regions or special status areas in other states such as Newfoundland. By such a study the advantages and disadvantages of independence might be clarified.[1] A study of the present institutional arrangement in developed small states such as Iceland and Luxembourg could also help to build a better institutional framework that suits all small states.

International cooperation between small states will certainly be beneficial although they are very diverse in terms of economic development, culture, political organisation and geographical position. There are common interests such as commitment to free international trade, interest in strengthening international law and order, protection of the environment, and a sustainable exploitation of the resources of the sea. It is also important to employ a common conservative policy when it comes to the question of revising the legal status of the nation-state. A big increase in the number of international actors could erode the respectable status of small states and the security of their legal position in the present global system.

To conclude, the experience of small states, particularly that of Iceland reported here, shows that traditional assumptions about the minimum size of states are not justified in economic and political terms. Small states have the capacity to provide a high level of welfare for their citizens and function successfully in the global system.

Note

1   A comparative study of North Atlantic islands has been initiated by the Institute of Island Studies, University of Prince Edward Island. The North Atlantic Islands Research Programme or Lessons From The Edge will collaborate with NordREFO, the research arm of the Nordic Council, to include Iceland, Greenland, Aland and the Faeroe Islands in the study along with Newfoundland, Prince Edward Island and the Isle of Man. See the Institute of Island Studies (1994).

Bibliography

Alþingi [The Icelandic Parliament] (1992), Frumvarp til laga um Seðlabanka Íslands [Draft of Central Bank Act], Reykjavík.

Arrow, Kenneth J. (1974), The Limits of Organizations, W. W. Norton & Company: New York.

Attard, David (1989), The Exclusive Economic Zone In International Law, Clarendon Press: Oxford.

Balassa, Bela (1973), *The Theory of Economic Integration*, George Allen & Unwin: London.

Barro , Robert J. and Xawier Sala-I-Martin (1991), 'Convergence across States and Regions', in William C. Brainard and George L. Perry (eds.), *Brookings Papers on Economic Activity*, Brookings Institution: Washington.

Björnsson, Ólafur (1967), 'Economic development in Iceland since World War II', *Weltwirtschaftliches Archiv*, Vol. 97, No. 2.

Björnsson, Ólafur (1994), 'Íslandsbanki níutíu ára' [The 90th anniversary of the Bank of Iceland], *Lesbók Morgunblaðsins*, Vol. 69, No. 24.

Blair, Patricia W. (1967), *The Ministate Dilemma*, Carnegie Endowment for International Peace: New York.

Blazic-Metzner, Boris and Helen Huges (1982), 'Growth Experience of Small Economies', in Bimal, Jalan (ed.), *Problems and Policies in Small Economies*, Croom Helm: London.

Blejer, Mario I. (1988), 'Regional Integration in Latin America: Experience and Outlook', *Journal of International Economic Integration*, Vol. 3, No. 2.

Borgese, E. M. and N. Ginsburg (eds.) (1982), *The Ocean Yearbook 3*, The University of Chicago Press: Chicago.

Bowen, Harry P., Edward E. Learner and Leo Sveikauskas (1977), 'Multicountry, Multifactor Tests of the Factor Abundance Theory', *American Economic Review*, December.

Brown, Lester R. et al. (1992), *State of the World 1992, A Worldwatch Institute Report on Progress Toward a Sustainable Society*, W. W. Norton & Company: New York.

Buchanan, Allen (1992), 'Moral Questions of Secession and Self-Determination', *Journal of International Affairs*, Vol. 45, No. 2.

Byggðastofnun [Institute of Regional Development] (1987), *Fríiðnaðarsvæði á Suðurnesjum* [Free Industrial Zone in the Suðurnes Region], Reykjavík.

Calvocoressi, Peter (1991), *World Politics Since 1945*, Longman: London.

Cleland, John G. and Susheela Singh (1980), 'Islands and the Demographic Transition', *World Development*, Vol. 8, No. 12, Pergamon Press: Oxford.

Cooper, James M. (1993), 'The Challenge To The Nation-State In International Law' [Paper presented to the 16th World Congress on Philosophy of Law and Social Philosophy], Reykjavík.

Dahl, Robert A. and Edward R. Tufte (1974), *Size and Democracy*, Stanford University Press: California.

Darwent D. E. (1969), 'Growth Poles and Growth Centres in Regional Planning, a Review', *Environment and Planning*, No. 1.

Deardorff, Alan V. (1984), 'Testing Trade Theories and predicting trade flows', in R. W. Jones and P.B. Kenen (eds.), *Handbook of International Economics*, Vol. 1, North Holland: Amsterdam.

Derbyshire, J. Denis and Ian Derbyshire (1991), *Spotlight on World Political Systems, an Introduction to Comparative Government*, Chambers: Edinburgh.

Deutsch, Karl W. (1967), 'Communication Models and Decision Systems', in James C. Charlesworth (ed.), *Contemporary Political Analysis*, The Free Press: New York.

Dolman, Antony J. (1985), 'Paradise lost', in Edward Dommen and Philippe Hein (eds.), *States, Microstates and Islands*, Croom Helm: London.

Dommen, Edward (1985), 'What is a microstate', in Edward Dommen and Philippe Hein (eds.), *States, Microstates and Islands*, Croom Helm: London.

Dommen, Edward and Philippe Hein (1985), 'Foreign trade in goods and services: The dominant activity of small island economies' in Edward Dommen and Philippe Hein (eds.), *States, Microstates and Islands*, Croom Helm: London.

Doumenge, François (1985), 'The viability of small intertropical islands', in Edward Dommen and Philippe Hein (eds.), *States, Microstates and Islands*, Croom Helm: London.

Easton, David (1965), *A Framework for Political Analysis*, Prentice-Hall: New Jersey.

Einstein, Albert (1973), *Ideas and Opinions*, Souvenir Press: London.

Elliott, Dorinda and Betsy McKay (1995), 'Out of Control', *Newsweek* , January 9, pp. 8–12.

Emerson, Machael et al. (1988), *The Economics of 1992: The E. C. Commission's Assessment of the Economic Effects of Completing the Internal Market*, Oxford University Press: Oxford.

Fiskifélag Íslands (1991), *Útvegur 1990* [Fisheries 1990], Reykjavík.

Ford, John (1963), 'The Ohlin-Heckscher Theory of the Basis of Commodity Trade', *The Economic Journal*, September, pp. 458–476.

Framkvæmdastofnun Ríkisins [The Economic Development Institute] (1982), *Hagtölur Landshluta, Suðurnes* [Regional Statistics, Suðurnes], Reykjavík.

Friedman, Milton (1968), 'The Methodology Of Positive Economics', in May Broadbeck (ed.), *Readings in the Philosophy of the Social Sciences*, Macmillan: London.

GATT (1990), *International Trade 89–90*, Vol. 1, Geneva.

Gíslason, Gylfi Þ, (1979), 'Ísland, Fríverzlunarsamtökin og Efnahagsbandalagið' [Iceland, The Free Trade Area and The Common Market], *Fjármálatíðindi* [The Journal of Finance], Vol. 16, No. 3, pp. 178–201.

Greenaway, David and Chris Milner (1986), *The Economics of Intra-Industry Trade*, Basil Blackwell: Oxford.

Grubel, H. G. and P. J. Lloyd (1975), *Intra-Industry Trade, The Theory and Measurement of International Trade in Differentiated Products*, Macmillan: London.

Hagfræðideild Landsbanka Íslands [Department of economics in Landsbanki of Iceland] (1960), 'Efnahagsáhrif varnarliðsins' [The Economic Impact of the Defence Force], *Fjármálatíðindi*, Vol. 7, No. 2 [Continued by Seðlabanki Íslands from 1962], pp. 83–88.

Hagfræðistofnun Háskóla Íslands [The Institute of Economics, University of Iceland] (1991a), *Efnahagssamvinna Evrópuþjóða og Hagstjórn á Íslandi* [European Economic Cooperation and Economic Policy in Iceland], Reykjavík [Translation based on unpublished English summary of main points from the National Economic Institute].

Hagfræðistofnun Háskóla Íslands (1991b), *Íslenskur Þjóðarbúskapur og Evrópska Efnahagssvæðið* [The Icelandic Economy and the European Economic Area], Reykjavík.

Hagstofa Islands [The Statistical Bureau of Iceland] (1978), *Hagtíðindi - Monthly Statistics*, Vol. 63, Reykjavík.

Hagstofa Islands (1984), *Tölfræðihandbók 1984. Statistical Abstract of Iceland 1984*, Reykjavík.

Hagstofa Islands (1991), *Landshagir 1991. Statistical Abstract of Iceland*, Reykjavík.

Hagstofa Islands (1992), *Hagtíðindi - Monthly Statistics*, Vol. 77, Reykjavík.

Hagstofa Islands (1993), *Landshagir 1993. Statistical Abstract of Iceland*, Reykjavík.

Handel, Michael (1990), *Weak States In The International System*, Frank Cass: London.

Haralz, Jónas H. (1962), 'Ísland og Efnahagsbandalag Evrópu frá efnahagslegu sjónarmiði séð' [Iceland and the European Economic Community from an economic point of view], *Fjármálatíðindi* [The Journal of Finance], Vol. 9, No. 1, pp. 7–23.

Hein, Philippe (1985), 'The Study Of Microstates', in Edward Dommen and Philippe Hein (eds.), *States, Microstates and Islands*, Croom Helm: London.

Held, David (1991), 'Democracy and the Global System', in David Held (ed.), *Political Theory Today* , Stanford University Press: California.

Helleiner, G. K. (1982), 'Balance of Payments Problems and Macro-economic Policy in Small Economies' in Bimal, Jalan (ed.), *Problems and Policies in Small Economies*, Croom Helm: London.

Henderson, Alexander M. (1969), 'The pricing of public utility undertakings' in Kenneth J. Arrow and Tibor Scitovsky (eds.), *Readings in Welfare Economics*, George Allen & Unwin: London.

Hirchman, Albert O. (1970), 'Interregional and International Transmission of Economic Growth', in McKee, David L., R. D. Dean and W. H. Leahy (eds.), *Regional Economics: Theory and Practice*, The Free Press: New York.

Hirsch, Seev (1967), *Location of Industry and International Competitiveness*, Oxford University Press: Oxford.

Hoffman, Mark S. (ed.) (1992), *The World Almanac and Book Of Facts 1993*, Pharos Books: New York.

Ingvarsson, Garðar (1987), 'Power Intensive Industries', in Jóhannes Nordal and Valdimar Kristinsson (eds.), *Iceland 1986*, The Central Bank of Iceland: Reykjavík.

Ingþórsson, Ágúst Þór (1993), 'Democracy and Bureaucracy and the Marginalization of the State' [Paper presented to the 16th World Congress on Philosophy of Law and Social Philosophy], Reykjavík.

International Monetary Fund (1988), *International Financial Statistics, Yearbook 1988*, New York.

Jackson, Peter (1992), 'Welfare Economics', in John Maloney (ed.), *What's new in Economics?*, Manchester University Press: Manchester.

Jalan, Bimal (1982), 'Classification of Economies by Size', in Bimal, Jalan (ed.), *Problems and Policies in Small Economies*, Croom Helm: London.

Jones, R. W. and J. P. Neary (1984), 'Positive Theory of International Trade', in R. W. Jones and P. B. Kenen (eds.), *Handbook of International Economics*, Vol. 1, North Holland: Amsterdam.

Jónsson, Eðvarð T. (1994), *Hlutskipti Færeyja* [The fate of the Faeroe Islands], Mál og Menning: Reykjavík.

159

Jónsson, Sigfús (1980), *The Development of the Icelandic Fishing Industry 1900–1940 and its Regional Implications*, The Economic Development Institute: Reykjavík.

Jovanovic, Miroslav N. (1992), *International Economic Integration*, Routledge: London.

Kaldor, Nicholas (1970), 'The Case for Regional Policies', *Scottish Journal of Political Economy*, Vol. 17, November.

Kaldor, Nicholas (1975), 'What is Wrong with Economic Theory', *The Quarterly Journal of Economics*, August.

Kaplan, Abraham (1964), *The Conduct of Inquiry, Methodology for Behavioural Science*, Chandler Publishing Company: Scranton.

Karlsson, Mikael M. (1992), 'Smáræða' [A short speech]. *Skírnir*, Vol. 166, Reykjavík, pp. 418–423.

Kennedy, Paul (1993), *Preparing for the Twenty-First Century*, Fontana Press: London.

Khalaf, Nadim G. (1971), *Economic Implications of the size of Nations with Special Reference to Lebanon*, E. J. Brill: Leiden.

Khatkhate, Deena R. and Brock K. Short (1980), 'Monetary and Central Banking Problems of Mini States', *World Development*, Vol. 8, No. 12, Pergamon Press: Oxford, pp. 1017–1025.

Kindleberger, Charles. P. and Peter. H. Lindert (1978), *International Economics*, Richard D. Irwin: Homewood.

Knox, A. D. (1967), 'Some Economic Problems of Small Countries', in Burton Benedict (ed.), *Problems of Smaller Territories*, Athlone Press: London.

Krugman, Paul R. (1990), *Rethinking International Trade*, The MIT Press: Cambridge, Mass.

Krugman, Paul R. (1991), *Iceland's Exchange Rate Regime: Policy Options* [A special report for the National Economic Institute and The Central Bank of Iceland], Reykjavík.

Kuznets, S. (1960), 'Economic Growth of Small Nations', in E. A. G. Robinson (ed.), *Economic Consequences of the Size of Nations*, Macmillan: London.

Kwiatkowska, Barbara (1989), *The 200 Mile Exclusive Economic Zone in the New Law of the Sea*, M. Nijhoft: Dordrech.

Lall, S. and S. Ghosh (1982), 'The Role of Foreign Investment and Exports in Industrialisation', in Bimal Jalan (ed.), *Problems and Policies in Small Economies*, Croom Helm: London.

Lewis, W. Arthur (1952), 'World Production and trade 1870–1960', *Manchester School*, May.

Linder, Staffan B. (1961), *An Essay on Trade and Transformation*, Wiley: New York.

Lipsey, R. G. and K. Lancaster (1973), 'The General Theory of Second Best', in M. J. Farrell (ed.), *Readings in Welfare Economics, A Selection of Papers from the Review of Economic Studies*, Macmillan: London.

Lloyd, P. J. and R. M. Sundrum (1982), 'Characteristics of Small Economies', in Bimal Jalan (ed.), *Problems and Policies in Small Economies*, Croom Helm: London.

Lloyd, Peter J. (1968), *International Trade Problems of Small Nations*, Duke University Press: Durham.

Madison, James (1988), 'Federalist No. 10', in Michael B. Levy (ed.), *Political Thought In America, An Anthology*, The Dorsey Press: Chicago.

Magnússon, Guðmundur (1991), *Some "invited" comments on Professor Krugman´s paper on "Iceland's Exchange Rate Regime: Policy Options"*, [unpublished paper for the National Economic Institute], Reykjavík.

Mál og Menning (1987), *The Times Atlas of the World*, Reykjavík.

Meadows, Donella H., Dennis L. Meadows, Jørgen Randers and William W. Behrens III (1972), *The Limits To Growth*, Universe Books: New York.

Meadows, Donella H., Dennis L. Meadows and Jørgen Randers (1992), *Beyond the Limits, Global Collapse or a Sustainable Future*, Earthscan Publications: London.

Meyer, Frederick Victor (1978), *International Trade Theory*, Croom Helm: London.

Meyer, Frederick Victor (1979), 'Intra-Industry Trade', *Aussenwirtschaft*, No. 4, pp. 324–340.

Mikhaílov, Nikolaj N. (1962), *Sovétríkin* [The Soviet Union], translated by Gísli Ólafsson, Heimskringla: Reykjavík.

Molinero, M. Rodríguez (1993), *The Principle of Subsidiarity* [Paper presented to the 16th World Congress on Philosophy of Law and Social Philosophy, Reykjavík].

Morgenthau, Hans J. (1968), *Politics Among Nations, The struggle for Power and Peace*, Alfred A. Knopf: New York.

Mundell, R. (1961), 'The Theory of Optimum Currency Areas', *American Economic Review*, September, pp. 657–665.

Myrdal, Gunnar (1957), *Economic Theory and Underdeveloped Regions*, Gerald Duckworth: London.

National Economic Institute (1980), *The Icelandic Economy, Developments 1979–1980*, Reykjavík.

Nicholson, Walter (1978), *Microeconomic Theory, Basic Principles and Extensions*, The Dryden Press: Hinsdale.

OECD (1974), *OECD Economic Surveys 1974, Iceland*, OECD Publications: Paris.

OECD (1990), *OECD Economic Surveys 1990, Iceland*, OECD Publications: Paris.

OECD (1994), *OECD Economic Surveys 1994, Iceland,* OECD Publications: Paris.

Ohlin, Bertil (1967), *Interregional and International Trade,* Harvard University Press: Cambridge, Mass.

Ohmae, K. (1993), 'The Rise of the Region State', *Foreign Affairs,* Vol. 72, No. 2, pp. 78–88.

Ólafsson, Björn Gunnar (1991), 'Byggðastefna, velferð og réttlæti' [Regional policy, welfare and justice], *Fjármálatíðindi* [The Journal of Finance], Vol. 38, No. 1, Reykjavík, pp. 49–56.

Ólafsson, Björn Gunnar (1993a), 'Gengisfelling leysir ekki vandann' [Devaluation will not solve the problem],*Vísbending* [A weakly bulletin on commerce and economics], Vol. 11, No. 12.

Ólafsson, Björn Gunnar (1993b), 'Trúverðug gengisstefna' [A credible exchange rate policy], *Vísbending,* Vol. 11. No. 37.

Perroux, François (1970), 'Note on the Concept of "Growth Poles"', in McKee, David L., R. D. Dean and W. H. Leahy (eds.), *Regional Economics: Theory and Practice,* The Free Press: New York.

Porter, Gareth and Janet Welsh Brown (1991), *Global Environmental Politics,* Westview Press: Boulder.

Porter, Michael E. (1990), *The Competitive Advantage of Nations,* The Macmillan Press: London.

Ragnarsson, Emil (1990), 'Orkunotkun við fiskveiðar. Þróun flota, sóknar, olíunotkunar og afla' [Use of energy in fishing. The development of fleet, fishing effort, use of gasoline and catches], *Ægir,* Vol. 83, No. 7.

Rapaport, Jacques, Earnst Muteba and Joseph J. Therattil (1971), *Small States & Territories: Status and problems,* Arno Press: New York.

Salvatore, Dominic (1990), *International Economics,* New York: Macmillan, 3rd edition.

Salz, Pavel (1993), *Regional, Socio-Economic Studies in the Fisheries Sector. Summary Report,* Commission of the European Communities: Luxembourg.

Sartori, Giovanni (1965), *Democratic Theory,* Frederick A. Praege: New York.

Schumacher, Friedrich Earnst (1989), *Small Is Beautiful: Economics as if People Mattered,* Harper & Row: New York.

Scitovsky, Tibor (1960), 'International Trade and Economic Integration as a Means of Overcoming the Disadvantages of a Small Nation', in E. A. G. Robinson (ed.), *Economic Consequences of the Size of Nations,* Macmillan: London.

Scitovsky, Tibor (1969), 'Two Concepts of External Economies', in Kenneth J. Arrow and Tibor Scitovsky (eds.), *Readings in Welfare Economics,* George Allen & Unwin: London.

Seðlabanki Íslands [The Central Bank of Iceland], *Hagtölur Mánaðarins Desember 1994* [Monthly Bulletin of Statistics], Reykjavík.

Sigurðsson, Jón (1970), 'Aðild Íslands að EFTA og fjárhagsmálin' [The Membership of Iceland in EFTA and Financial Matters], *Fjármálatíðindi* [The Journal of Finance], Vol. 17, No. 1.

Sinjela, Mpazi A. (1989), 'Land-Locked States Rights in the Exclusive Economic Zone from the Perspective of the UN Convention on the Law of the Sea: An Historical Evaluation', *Ocean Development and International Law*, Vol. 20, No. 1.

Smith, Brian. C. and Jeffrey Stanyer (1980), *Administering Britain*, Martin Robertson: Oxford.

Sprout, Ronald V. A. and James H. Weaver (1992), 'International Distribution of Income: 1960–1987', *KYKLOS*, Vol. 45, pp. 237–258.

Sverrisson, Hörður (1984), 'Verðtrygging í viðskiptabönkum. Reikningar innlánsstofnana 1981 og 1982' [Indexing in commercial banks. Accounts of deposit money institutes 1981 and 1982], *Fjármálatíðindi*, Vol. 31, No. 1.

The Institute of Islands Studies (1994), *Lessons From The Edge, The Newsletter of the North Atlantic Islands Programme*, Vol. 1, No. 1.

The World Bank (1990a), *World Tables 1989–1990 Edition*, Baltimore.

The World Bank (1990b), *World Development Report 1990*, Oxford.

The World Bank (1992), *Trends in Developing Economies 1992*, Washington.

The World Bank (1994), *World Development Report 1994*, Oxford.

Tolley, George and John Crihfield (1987), in Edwin S. Mills (ed.), *Handbook of Regional and Urban Economics, Volume 2 Urban Economics*, North-Holland: Amsterdam, pp. 1285–1311.

Tullock, Gordon (1969), 'Federalism: Problems Of Scale', *Public Choice*, Vol. 6, Spring.

UNCTAD (1988), *Trade and Development Report 1988*, United Nations: New York.

UNCTAD (1991), *Handbook of International Trade and Development 1990*, United Nations: New York.

United Nations (1991), *International Trade Statistics Yearbook 1989*, New York.

United Nations (1992), *Statistical Yearbook 1988/89*, New York.

United Nations (1993), *Population and Vital Statistics Report*, UN Statistical Papers, Series A, Vol. 15, No. 1, New York.

Utanríkisráðuneytið [Ministry of Foreign Affairs] (1992), *EES - samningurinn Sjávarútvegsmál* [The EEA Treaty - Fisheries], Reykjavík.

Vital, David (1967), *The Inequality of States: A Study of the Small Power in International Affairs*, Clarendon Press: Oxford.

Waddell, D. A. G. (1967), 'Case Study: British Honduras', in Burton Benedict (ed.), *Problems of Smaller Territories*, The Athlone Press: London.

Ward, R. G. (1967), 'The Consequences of Smallness in Polynesia', in Burton Benedict (ed.), *Problems of Smaller Territories*, The Athlone Press: London.

Young, Allyn A. (1969), 'Increasing Returns and Economic Progress', in Kenneth J. Arrow and Tibor Scitovsky (eds.), *Readings in Welfare Economics*, George Allen & Unwin: London.

Þjóðhagsstofnun [The National Economic Institute] (1977), Hagur iðnaðar og aðild Íslands að Fríverslunarsamtökum Evrópu og viðskiptasamningur Íslands við Efnahagsbandalag Evrópu [The Prospects of Industry and the Association of Iceland with EFTA and the Trade Treaty of Iceland and the EC], Reykjavík.

Þjóðhagsstofnun (1991a), Sögulegt Yfirlit Hagtalna 1945–1988 [Historical Statistics], Reykjavík.

Þjóðhagsstofnun (1991b), *Búskapur hins opinbera 1980–1988* [General Government 1980–1988], Reykjavík.

Þjóðhagsstofnun (1991c), *Áhrif Innri Markaðar Evrópubandalagsins á Íslanskan Iðnað* [The Impact of the Internal Market of the European Community on Icelandic Industry], Reykjavík.

Þjóðhagsstofnun (1993), Sögulegt Yfirlit Hagtalna, Historical statistics, Reykjavík.

Þjóðhagsstofnun (1994), *Atvinnuvegaskýrsla 1991* [Industrial Statistics 1991], Reykjavík.